SpringerBriefs in Electrical and Computer Engineering

Control, Automation and Robotics

Series editors

Tamer Başar
Antonio Bicchi
Miroslav Krstic

More information about this series at http://www.springer.com/series/10198

Bruce A. Francis · Manfredi Maggiore

Flocking and Rendezvous in Distributed Robotics

 Springer

Bruce A. Francis
Department of Electrical and Computer
  Engineering
University of Toronto
Toronto, ON
Canada

Manfredi Maggiore
Department of Electrical and Computer
  Engineering
University of Toronto
Toronto, ON
Canada

ISSN 2191-8112   ISSN 2191-8120   (electronic)
SpringerBriefs in Electrical and Computer Engineering
ISSN 2192-6786   ISSN 2192-6794   (electronic)
SpringerBriefs in Control, Automation and Robotics
ISBN 978-3-319-24727-4   ISBN 978-3-319-24729-8   (eBook)
DOI 10.1007/978-3-319-24729-8

Library of Congress Control Number: 2015950026

Springer Cham Heidelberg New York Dordrecht London

Printed on acid-free paper

Springer International Publishing AG Switzerland is part of Springer Science+Business Media
(www.springer.com)

To Lian Francis

Preface

Imagine a flock of birds migrating south before winter. The flock is composed of identical birds, but no one bird is the leader, nor can any one bird always see all the other birds. The birds obviously have a common objective—to fly from one general area to another specified location. At any one time there is a small number of birds at the front, but this group changes as the birds fatigue. The birds cooperate in order to fulfill the overall function of group migration. We can view this flock of birds as a *distributed system*, the adjective "distributed" not necessarily meaning geographically distributed, but rather distributed in function or authority. Besides a flock of birds you could imagine a school of fish or a colony of ants. Now think of such a system except where each bird (or fish or ant) is replaced by a mobile robot: a wheeled rover on the ground or a quadrocopter in the air. Then you have a distributed system of robots, and that is the subject of this book.

In the late twentieth century the research subject of distributed robotics burst on the scene, grew very quickly, and has become a core subject within control theory and engineering. Computer scientists, control theorists, roboticists, and other scientists and engineers have contributed to the subject. There is now a significant body of theory and also of experimental prototypes.

This book takes as its launchpad the 2014 IEEE Bode Lecture entitled "The Rendezvous Problem." The book covers the two most basic problems of distributed robotics, the flocking problem and the rendezvous problem, for wheeled robots and quadrocopters. Quadrocopters was not touched in the Bode Lecture and consequently is a feature of this book. The book is aimed at graduate students and others who wish to get into this subject. We view formation flying of quadrocopters as an especially fertile field for new Ph.D. theses.

Acknowledgments

It is our pleasure to thank the students who worked on the subject of distributed robotics under our supervisions: Alfred Sum, Zhiyun Lin, Joshua Marshall, Stephen Smith, Laura Krick, Hien Goi, Flörian Dörfler, Johannes Dold, Mohammed

El-Hawwary, and Ashton Roza. We also thank Mireille Broucke, Tim Barfoot, and Luca Scardovi, our Toronto colleagues with whom we co-supervised some of these students. Finally, thanks to Avraham Feintuch for his collaboration on the subject of infinitely many robots (Sect. 4.6).

Contents

Chapter 1
Introduction

1.1 Motivation

The problem of coordinated control of a network of mobile autonomous robots is
of interest in control and robotics because of the broad range of potential applications: planetary exploration, operations in hazardous environments, games such as
robot soccer, and so on. Distributed robot networks can potentially exhibit structural
flexibility, reliability through redundancy, and simple hardware as compared to a
complex individual robot.

The first robot rover exploration of Mars was in 1997—the Mars Pathfinder Mis
sion. The rover, named Sojourner, is shown in Fig. 1.1. You can see the whip antenna.
The radio link was used to send commands from Earth to the rover and receive images
and other data from the rover. Because the rover radio had a signal range similar to a
walkie-talkie, namely, about 10 m, all rover communication was done with the aid of
the lander communications interface, as in Fig. 1.2. The rover telecommunications
system was a two-way wireless UHF (Ultra High Frequency) radio link between the
lander and the rover. The rover's and lander's UHF antennas worked very much like
the antennas on walkie talkies or on car radios, using a "monopole" antenna. The
signal to be transmitted enters the antenna through a coaxial connector located at the
bottom, travels through a short section of balanced coaxial line, and is radiated by
the monopole.

It is desirable to have an antenna radiation pattern shaped to match its particular
application. Satellite dishes are designed to look at a particular location in space and
therefore need to have narrow and directive radiation patterns. The rover antenna did
not need to look up into space, but rather needed to look horizontally in 360° given
that the lander could be in any direction. An ideal monopole has a 360° radiation
pattern that is donut shaped, oriented horizontally. It is not meant to look straight
up, and has poor reception in that direction. Certain metallic or rocky structures and
ground reflections near the monopole antenna will distort its radiation pattern and
cause holes or null zones to form. In these null zones the signal can drop significantly,

© The Author(s) 2016
B.A. Francis and M. Maggiore, *Flocking and Rendezvous in Distributed
Robotics*, SpringerBriefs in Control, Automation and Robotics,
DOI 10.1007/978-3-319-24729-8_1

Fig. 1.1 Sojourner (Jet
Propulsion Lab, NASA).
(This image is in the public
domain and was downloaded
from the Web)

Fig. 1.2 Communications
interface (NASA). (This
image is in the public
domain and was downloaded
from the Web)

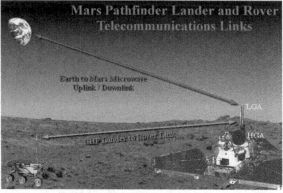

causing poor reception. It is important to know where the rover is relative to the lander
when these null zones exist, for if two nulls happened to get pointed at each other,
there may be no radio reception at all.

  As discussed, Sojourner had to be within about 10 m of the lander to send radio
signals. This obviously is a limitation for scientific experiments. To get a longer radio
link, one could use a higher power signal. But power on a Mars robot is a luxury.
Another solution is an antenna array.

Fig. 1.3 High Frequency Active Auroral Research Program. (This image is in the public domain and was downloaded from the Web)

The purpose of an antenna array is to achieve directivity, the ability to send the transmitted signal in a preferred direction. If a large number of array elements can be used, it is possible to greatly enhance the strength of the signal transmitted in a given direction. An interesting example is the High Frequency Active Auroral Research Program (HAARP) in Alaska, whose purpose is to study the ionosphere. The site consists of a 15×12 array of dipole antennas: see Fig. 1.3.

This leads us to the following scenario. A team of rovers doing scientific experiments. Each has, besides scientific instruments, a radio transceiver and an antenna. When it is time to communicate with the lander, the rovers arrange themselves in a suitable formation to become an antenna array in order to optimize the signal strength. In general, the larger the array, the higher the resolution it can achieve.

The preceding example leads to the sub-question: can we get a group of robot rovers, placed initially at random, to form a circle or other shape? We study this question for the simplest possible model of a robot, a point moving in the plane, then for a model of a wheeled rover moving in the plane, and finally for a quadcopter in 3D space.

This monograph is about control theoretic robotics problems. There is frequently a hubbub about the gap between theory and practice. Let us be clear about that: Real problems cannot be solved just by applying formulas. So the methodology of control engineering is to begin with a real problem; to abstract the central issues and formulate an idealized, hypothetical problem; to develop, if necessary, new mathematical methods for its solution; and to work out a rigorous solution. Then one has a framework on which to do the real problem.

For more about robots in space, go to http://www.jpl.nasa.gov.

1.2 Models, Sensing, and Control Specifications

In this monograph we present the two most basic distributed robotics problems: flocking and rendezvous. The flocking problem is to get all the robots to move in the same direction at the same speed; the rendezvous problem is to get all the robots to converge to the same meeting point. Such objectives are achieved by, possibly among other factors, interacting with other robots. We call these others **neighbours**. Besides the two objectives of flocking and rendezvous, one may characterize the setup by how neighbours are defined: by proximity or just fixed from the start. For example, one may distinguish n robots by numbering them and displaying the numbers on them. Then the neighbour structure could be sequential, like this: 1's neighbour is 2, 2's neighbour is 3; etc.; n's neighbour is 1. This is called **cyclic pursuit** and is an example of a fixed neighbour structure. Alternatively, the neighbours of robot i could be all other robots within, say, d metres of robot i. This is an example of a proximity-based neighbour structure.

We see from the discussion above that there are three dimensions to consider when classifying distributed robotics problems: the model of the robot one wishes to control (e.g., the unicycle model); the sensing constraint (what sensors are available, and who can see whom at any given time); and the control specification (e.g., rendezvous). In this book we focus on three model classes: integrator points, kinematic unicycles, and flying vehicles. We present two types of sensing constraints:

Fig. 1.4 Twelve problems:
the goal can be flocking or
rendezvous; the robots can
be integrator points,
unicycles, or flying vehicles;
the neighbour sets can be
fixed or proximity dependent

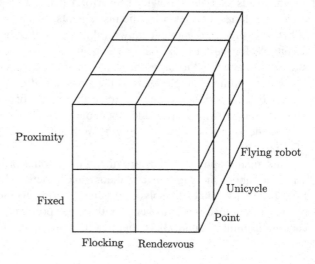

fixed neighbour structure and proximity-based neighbour structure. Finally, we investigate two control specifications: flocking and rendezvous. In this way we have $2 \times 2 \times 3 = 12$ problems (Fig. 1.4).

It will turn out that not all twelve problems make sense, since flocking is a degenerate problem for integrator points. Moreover, many of these problems are as yet open. In the case of a proximity-based neighbour structure, flocking is an open problem for all model classes, and rendezvous has been solved only for integrator points. In the case of a fixed neighbour structure, the flocking problem has been solved for both unicycles and flying vehicles, while the rendezvous problem has been solved only for integrator points and unicycles.

1.3 Notation

The notation follows fairly standard conventions in signals and systems. The set of integers—negative, zero, and positive—is denoted $\mathbb{Z}$. Continuous time and discrete time are both denoted by t; context will determine whether t is a real number ($t \in \mathbb{R}$) or an integer ($t \in \mathbb{Z}$). Dot, as for example $\dot{x}$, denotes derivative with respect to time t.

A vector in $\mathbb{R}^2$ is written as an ordered pair

$$x = (x_1, x_2)$$

or as a column vector

$$x = \begin{bmatrix} x_1 \\ x_2 \end{bmatrix},$$

whichever is more convenient at the time. Likewise, we might associate

$$(u, v) \text{ and } \begin{bmatrix} u \\ v \end{bmatrix}$$

where u is an m-tuple and v an n-tuple. This permits us to avoid ugly expressions like $\begin{bmatrix} u^T & v^T \end{bmatrix}^T$ for a column vector.

Mathematically, we can regard the plane as being the Euclidean plane $\mathbb{R}^2$ or the complex plane $\mathbb{C}$. If $x = (x_1, x_2)$, $y = (y_1, y_2)$ are two real vectors in $\mathbb{R}^2$, their dot product is written

$$\langle x, y \rangle = x_1 y_1 + x_2 y_2.$$

Likewise and equivalently, members of $\mathbb{C}$ are written $x_1 + jx_2$. If $x = x_1 + jx_2$ and $y = y_1 + jy_2$, to be consistent with $\mathbb{R}^2$ the dot product of x and y is defined to be (overbar denotes complex conjugate and Re denotes real part)

$$\begin{aligned}
\langle x, y \rangle &= \operatorname{Re} x\overline{y} \\
&= \operatorname{Re} (x_1 + jx_2)(y_1 - jy_2) \\
&= x_1 y_1 + x_2 y_2 \\
&= \operatorname{Re} \overline{x} y.
\end{aligned}$$

Finally, we let $\{e_1, \ldots, e_n\}$ denote the natural basis of $\mathbb{R}^n$. Thus the vector e_i has a one in the ith position and all other elements are zero.

Chapter 2
Models of Mobile Robots in the Plane

2.1 The Common Models

In this book all systems, including robots, are modeled as operating in continuous time. This is the natural world given to us by Newtonian physics. Controllers are continuous-time too, but can be implemented digitally with samplers.

2.1.1 A 1D Rover

We begin with the simplest example: a wheeled rover of unit mass moves along a straight infinite road that runs through the 2D plane. We can take the plane to be either $\mathbb{C}$ or $\mathbb{R}^2$; we take the former for now. By translating and rotating if necessary, we may suppose the road is the real line (the horizontal axis). The position of the rover on the road is denoted by the real variable z. The rover has an onboard motor that drives a wheel without slipping, imparting f Newtons of force (negative f implies the force is to the left). We neglect viscous friction and say that Newton's second law is applicable:

$$\ddot{z} = f.$$

Equivalently,

$$\dot{z} = v, \quad \dot{v} = f.$$

See Fig. 2.1. Furthermore, if the robot has a velocity sensor, a high-gain feedback in an inner loop, as shown in Fig. 2.2, converts the double integrator into the single

© The Author(s) 2016
B.A. Francis and M. Maggiore, *Flocking and Rendezvous in Distributed Robotics*, SpringerBriefs in Control, Automation and Robotics,
DOI 10.1007/978-3-319-24729-8_2

Fig. 2.1 The simplest rover.
Force input, position output

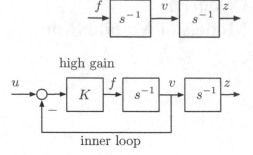

Fig. 2.2 A high-gain inner
loop. If K is large, from u to
z is approximately s^{-1}

integrator. A high-gain inner loop is placed around the dynamics. The transfer function from f to z is s^{-2}, but from u, a command velocity, to z is approximately just
s^{-1} if K is sufficiently large. Thus a high-gain loop around the dynamics gives

$$\dot{z} = u.$$

2.1.2 2D Integrator Point

Now we turn to 2D robot models. The first obvious generalization of the 1D rover is
a unit mass moving freely on the plane. The position z of the mass is now a complex
variable and so is the force f:

$$\ddot{z} = f.$$

Just as in the 1D case, we may use high-gain feedback and view the velocity as a
control input, in which case we obtain the model of an **integrator point**

$$\dot{z} = u.$$

Letting $z := x + jy$ and $u := v + jw$ we get an equivalent model in real variables:

$$\dot{x} = v$$
$$\dot{y} = w.$$

Is an integrator point a physically meaningful model of a robot? The robots we
consider are rigid bodies, or made up of rigid bodies. A rigid body in 3D has six
degrees of freedom (three for translation and three for rotation). A rigid body in 2D
has three degrees of freedom (two for translation and one for rotation). Integrator
points cannot represent something physical in the plane because there are only two
degrees of freedom instead of three. To account for the missing degree of freedom,
angular position, we could write

$$\dot{x} = v$$
$$\dot{y} = w$$
$$\dot{\theta} = 0.$$

This robot can move in any direction but its orientation does not change, and therefore it is not related to motion. While it is possible to devise a mechanism that decouples the orientation of the robot from its direction of motion, it is rather uncommon to find physical robots with this property. Many researchers use integrator points so that control problems can be easily solved, but it is not always clear how to apply such solutions to physical robots.

Realistic models of 2D robots couple the robot's orientation with its direction of motion. An example of this kind of robot is shown in Fig. 2.3. The robot has an omnidirectional camera (a conventional camera pointing up at a conical mirror), two wheels with independent motor drives, and a laptop to store a controller program. The robot is confined to move on a floor (it cannot fly). Thus as a mechanical dynamical system it has three degrees of freedom, which in conventional notation are x, y, θ. The vector (x, y) locates the centre of mass on the floor, and θ specifies the heading angle as measured from some fixed direction.

Fig. 2.3 A wheeled robot with an omnidirectional camera. (This image is in the public domain and was downloaded from Wikipedia, Omnidirectional camera)

Fig. 2.4 The unicycle

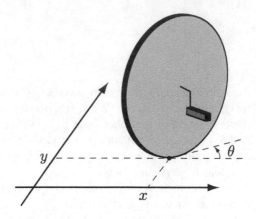

To model a robot of the kind represented in Fig. 2.3, we turn to unicycles and bicycles.[1] These are kinematic models. Of course, a real robot has dynamics too, but this can frequently be removed by a high-gain inner loop as we just did in Fig. 2.2. Sometimes, it is convenient to make the complex plane the workspace where the robots live. Recall that every complex number w can be written uniquely in polar form as $w = ve^{j\theta}$, where $v = |w|$ and θ is a real number in the interval $[0, 2\pi)$.

2.1.3 Unicycles

A kinematic unicycle is a robot with one steerable drive wheel (see Fig. 2.4). If we assume that the wheel is always perpendicular to the ground, we may represent the unicycle on the complex plane as in Fig. 2.5, where $z := x + jy$ is the position vector and $e^{j\theta}$ the normalized velocity vector.

Convention. In Fig. 2.5 we represent a complex number in two different ways. First, z is shown as a dot, obviously at the correct location in the complex plane. On the other hand, $e^{j\theta}$, which is a complex number too, is shown as an arrow. The convention is likely familiar. An element of $\mathbb{C}$ (or $\mathbb{R}^2$) can be regarded geometrically as a *geometric vector* or a *point*. A point, depicted as a dot, identifies a position in space. A geometric vector, depicted as an arrow, identifies a magnitude and a direction.

We return to the unicycle. Its degrees of freedom are x, y, θ, just as for the robot in Fig. 2.3. From $\dot{z} = ve^{j\theta}$ we get the two equations

---

[1]Of course in everyday parlance *bicycle* refers to a real physical two-wheel vehicle that people ride. We use the same word also for something else, namely, a mathematical model of the kinematic part of a real bicycle. Likewise for *unicycle*.

Fig. 2.5 Unicycle in the complex plane

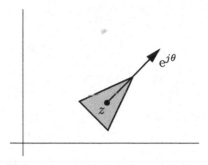

$$\dot{x} = v\cos(\theta)$$
$$\dot{y} = v\sin(\theta).$$

Defining $\omega = \dot{\theta}$ we get a third equation. In this way we arrive at the state equations

$$\dot{x} = v\cos(\theta)$$
$$\dot{y} = v\sin(\theta) \qquad (2.1)$$
$$\dot{\theta} = \omega.$$

The state variables are x, y, θ and the inputs are v, ω. In terms of complex variables we have

$$\dot{z} = ve^{j\theta}$$
$$\dot{\theta} = \omega.$$

There is a third equivalent model in which one regards the unicycle as a *moving orthonormal frame*. Consider the **body frame** $\mathcal{B} = \{r, s\}$ attached to the unicycle— see the picture on the left in Fig. 2.6. The origin of the frame is at (x, y); r is the normalized velocity vector, $r = e^{j\theta}$; finally, s is the counterclockwise rotation of r by $\pi/2$, $s := jr$. Thus

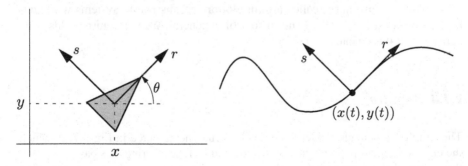

Fig. 2.6 Unicycle body frame and the Frenet–Serret frame of a regular *curve*

$$\dot{r} = \frac{d}{dt}e^{j\theta}$$
$$= je^{j\theta}\dot{\theta}$$
$$= jr\omega$$
$$= s\omega.$$

Likewise

$$\dot{s} = -r\omega.$$

Using (z, r, s) as state of the unicycle, we find that the state model is

$$\dot{z} = vr$$
$$\dot{r} = s\omega \qquad\qquad (2.2)$$
$$\dot{s} = -r\omega.$$

The control inputs are, as before, v and ω.

There is an intriguing relationship between the body frame $\mathcal{B}$ defined above and the **Frenet–Serret frame** of differential geometry [7]. In differential geometry, the Frenet–Serret frame is a moving orthonormal frame one associates with a regular curve—see the picture on the right in Fig. 2.6. The relationship between the frame $\mathcal{B}$ defined earlier and the Frenet–Serret frame is this: If in the unicycle model (2.2) we set $v(t) \equiv 1$ and we let $\omega(t)$ be an arbitrary continuous function, then the moving frame $\{r(t), s(t)\}$ is precisely the Frenet–Serret frame associated with the curve $z(t)$ traced by the unicycle on the complex plane. Moreover, $\omega(t)$ is the *signed curvature* of the curve. In differential geometry, the last two equations in (2.2) are called the *Frenet–Serret formulas* associated with the curve $z(t)$.

We conclude this part with a remark. The unicycle can move only in the direction it is heading. That is, there is a no side-slip condition. To derive it, note that the velocity vector of the unicycle is parallel to the body frame vector r, which in turn is perpendicular to the body frame vector s. In other words, $\langle \dot{z}, s \rangle = 0$, or

$$-\dot{x}\sin(\theta) + \dot{y}\cos(\theta) = 0.$$

This velocity constraint is called a **nonholonomic constraint**. Systems with non-holonomic constraints are difficult to control in general. We will return to this issue at the end of this chapter.

2.1.4 Bicycles

The simplest kinematic model of a bicycle is the one depicted in Fig. 2.7, in which the bicycle frame is perpendicular to the ground and the steering axis passes through the centre of the front wheel. We denote by (x, y) the coordinates of the point of

Fig. 2.7 Schematic: (x, y) is the location of the rear wheel, B is the wheelbase, θ is the angle of the frame with respect to the x-axis, γ is the angle of the front wheel with respect to the frame

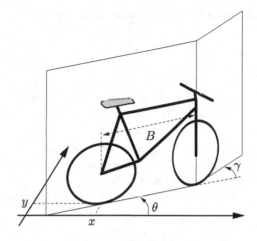

contact of the rear wheel with the ground. We let θ be the angle that the frame makes with the x axis, and γ the steering angle, as in the figure. While this model might not be a faithful representation of a real bicycle, it turns out to be quite useful because it captures the essential features of a car with four wheels, only the front two being steerable.

Since the bicycle is assumed to be perpendicular to the ground, we may represent it on the complex plane as in Fig. 2.8. In the figure, z_1 is the position of the rear wheel, i.e., $z_1 = x + jy$, and z_2 the position of the front wheel. The vector r_1 is the normalized difference $z_2 - z_1$, while r_2, also a unit vector, represents the heading of the front wheel. In terms of the angles θ and γ, we have $r_1 = e^{j\theta}$ and $r_2 = e^{j(\theta+\gamma)}$.

We see that the bicycle has four degrees of freedom: x, y, θ, γ. We take as control inputs the speed of the point z_1 and the steering rate $\dot{\gamma}$. We denote them by v and ω, respectively.

Fig. 2.8 Variables used to describe the bicycle: z_1 and z_2 are the positions of the two wheels; r_1 and r_2 are the normalized velocity vectors

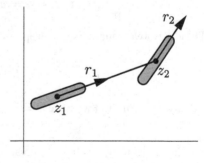

Lemma 2.1 *The kinematic model of the bicycle in Fig. 2.7 is*

$$\dot{x} = v\cos(\theta)$$
$$\dot{y} = v\sin(\theta)$$
$$\dot{\theta} = \frac{v}{B}\tan(\gamma) \tag{2.3}$$
$$\dot{\gamma} = \omega.$$

Proof The velocity of the point of contact z_1 is proportional to r_1 and its magnitude is v. Thus, $\dot{z}_1 = vr_1$. Writing the real and imaginary parts of this identity we get the first two equations in (2.3).

Letting $v_1 := v$ and $v_2 := |\dot{z}_2|$, we have the following equations:

$$r_1 = e^{j\theta}, \quad r_2 = e^{j(\theta+\gamma)}, \quad z_2 - z_1 = Br_1 \tag{2.4}$$

$$\dot{z}_1 = v_1 r_1, \quad \dot{z}_2 = v_2 r_2. \tag{2.5}$$

Differentiating the third equation in (2.4) we get

$$\dot{z}_2 - \dot{z}_1 = B\dot{r}_1.$$

Substitute from Eqs. (2.4) and (2.5):

$$v_2 r_2 - v_1 r_1 = Bj\dot{\theta}r_1.$$

Divide by r_1:

$$v_2 e^{j\gamma} - v_1 = Bj\dot{\theta}.$$

Write the real and imaginary parts:

$$v_2\cos(\gamma) = v_1, \quad v_2\sin(\gamma) = B\dot{\theta}.$$

Divide the two equations and drop the subscript on v_1:

$$\dot{\theta} = \frac{v}{B}\tan(\gamma).$$

Finally, by definition we have $\omega = \dot{\gamma}$. $\square$

The model (2.3) has four state variables, x, y, θ, γ, and two inputs, v, ω. Like the unicycle, the bicycle has a nonholonomic constraint, the no-side slip condition of the rear wheel. The constraint is

$$-\dot{x}\sin(\theta) + \dot{y}\cos(\theta) = 0.$$

Fig. 2.9 Block diagram of the bicycle model

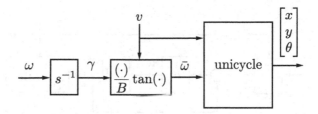

Compare the bicycle and unicycle models in (2.3) and (2.1). The two are very similar. A control law developed for the unicycle can be adapted, with some limitation, to the bicycle. The limitation is the obvious one that the front wheel of the bicycle must never become orthogonal to the rear wheel.

Now consider the block diagram of the bicycle model shown in Fig. 2.9. The middle box stands for the two-input, single-output nonlinear function $(\gamma, v) \mapsto \bar{\omega}$ given by

$$\bar{\omega} = \frac{v}{B} \tan(\gamma).$$

The unicycle may be regarded as a subsystem. If we place a high-gain inner loop around the dynamics of the steering angle, the bicycle and the unicycle become approximately equivalent. More precisely, let $v^\star(t) > 0$ and $\omega^\star(t)$ be arbitrary continuous signals. Define

$$\bar{\gamma}(t) := \arctan(B\omega^\star(t)/v^\star(t)),$$

and assume that $\bar{\gamma}(t)$ is a bounded signal.[2] Define the following control law for the steering rate of the bicycle:

$$\omega(t) = K(\bar{\gamma}(t) - \gamma(t)).$$

Here, $K > 0$ is a large gain. The block diagram of the bicycle with this control law[3] is depicted in Fig. 2.10. There is a high-gain negative feedback loop around the steering angle, so that we have the approximate identity $\gamma(t) \approx \bar{\gamma}(t)$. Assuming that this is true, from the block diagram in Fig. 2.10 we have

$$\bar{\omega}(t) = \frac{v^\star(t)}{B} \tan(\gamma(t)) \approx \frac{v^\star(t)}{B} \tan(\bar{\gamma}(t)) = \omega^\star(t).$$

Thus, in the block diagram of Fig. 2.10 we have the approximate identity $\bar{\omega}(t) \approx \omega^\star(t)$, and the closed-loop bicycle dynamics are approximately equivalent to the dynamics of a unicycle with control input $(v^\star, \omega^\star)$.

---

[2]Note that $\bar{\gamma}(t)$ may be unbounded if $v^\star(t) \to 0$ as $t \to \infty$.

[3]The signal $\bar{\gamma}(t)$ lies in the interval $(-\pi/2, \pi/2)$. If the steering angle γ is initialized in this interval, then it remains in it for all positive time, and therefore the signal $\bar{\omega}(t)$ is well-defined for all $t \geq 0$.

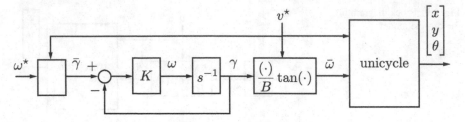

Fig. 2.10 A high-gain inner loop makes the bicycle look like a unicycle

In conclusion, if we are given control laws $(v^\star, \omega^\star)$ for the unicycle, we get control laws (v, ω) for the bicycle through the formulas

$$v = v^\star$$
$$\omega = K\left(\arctan(B\omega^\star/v^\star) - \gamma\right), \; K > 0 \text{ large}.$$

This works only if $v^\star(t) > 0$ and $\omega^\star(t)/v^\star(t)$ is bounded. In other words, if the speed of the unicycle tends to zero we require the angular speed of the unicycle to tend to zero at least as fast. This strategy, therefore, may be problematic when the control specification requires the bicycle to stop (such is the case in the solution to the rendezvous problem presented in Chap. 4).

We stress that the argument outlined above is not mathematically rigorous, which is why we do not state it as a theorem. A rigorous argument would rely on singular perturbation theory. Nonetheless, the argument suggests that if one can solve a control problem for the unicycle model, then it is possible to obtain a solution for the bicycle model. For this reason, in this monograph we focus our attention on the unicycle model.

2.1.5 Summary

1. The model of an integrator point robot on the complex plane is

$$\dot{z} = u.$$

In terms of real variables the model is

$$\dot{x} = v$$
$$\dot{y} = w.$$

This model is kinematic—mass is not included. It has only two degrees of freedom, x and y coordinates. It is not a complete model of a physical robot because its orientation is fixed. The reason point robots were introduced was so that control problems could be solved. How to apply these mathematical solutions is not always obvious.

2. A unicycle is a mathematical model of a wheeled robot with one steerable drive wheel. Again it is a kinematic model. It is a more realistic model of a mobile robot than is an integrator point robot. The equations of the unicycle on the real plane are

$$\dot{x} = v \cos(\theta)$$
$$\dot{y} = v \sin(\theta)$$
$$\dot{\theta} = \omega.$$

The model on the complex plane is

$$\dot{z} = v e^{j\theta}$$
$$\dot{\theta} = \omega.$$

We can also view the unicycle as a moving orthonormal frame, in which case its equations are

$$\dot{z} = vr$$
$$\dot{r} = s\omega$$
$$\dot{s} = -r\omega.$$

3. A bicycle is a mathematical model of a wheeled car-type robot, with a non-steerable drive wheel and a steerable non-drive wheel. It is a kinematic model. It has four degrees of freedom. The equations are

$$\dot{x} = v \cos(\theta)$$
$$\dot{y} = v \sin(\theta)$$
$$\dot{\theta} = \frac{v}{B} \tan(\gamma)$$
$$\dot{\gamma} = \omega.$$

The bicycle and unicycle models are very similar. A controller developed for the unicycle can be adapted to the bicycle, with some limitation.

2.2 Feedback Linearization of the Unicycle

As we emphasized in the previous section, the integrator point robot is not a good model of a real wheeled robot, whereas a unicycle is (except for articulated vehicles). The virtue of an integrator point robot is that it makes it easier to solve mathematical

Fig. 2.11 Feedback
linearization of a unicycle
about a point just ahead

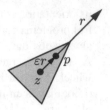

problems. However, a unicycle can be feedback linearized into an integrator point robot. This suggests that to solve a wheeled robot problem, one can first feedback linearize the unicycle robot, then solve the problem, and finally transform back to the wheeled robot. Will this work? We begin to look at this question in this brief section. We will return to it in Chap. 4.

Start with the unicycle model viewed as a moving orthonormal frame:

$$\dot{z} = vr$$
$$\dot{r} = s\omega$$
$$\dot{s} = -r\omega.$$

Let $\varepsilon > 0$. The point

$$p = z + \varepsilon r \tag{2.6}$$

is a distance ε in front of the unicycle, as shown in Fig. 2.11. Differentiate both sides of (2.6) with respect to t to get

$$\dot{p} = vr + \varepsilon s\omega.$$

Define u to be the right-hand side:

$$u := rv + \varepsilon s\omega. \tag{2.7}$$

Take inner products of both sides of this equation first with r and then with s. Since r, s are orthonormal, we obtain

$$v = \langle u, r \rangle, \quad \omega = \varepsilon^{-1} \langle u, s \rangle. \tag{2.8}$$

The dynamics of the point p are simply

$$\dot{p} = u. \tag{2.9}$$

To recap, the feedback linearized unicycle model is (2.9), which is in terms of the point just-ahead p; the input u is related to the physical inputs v, ω via Eqs. (2.7) and (2.8).

Fig. 2.12 A unicycle controlled to go to the origin: $\varepsilon = 0.1$, $(x(0), y(0)) = (-1, 1)$ and $(x(0), y(0)) = (1, 1)$. The axes are x and y

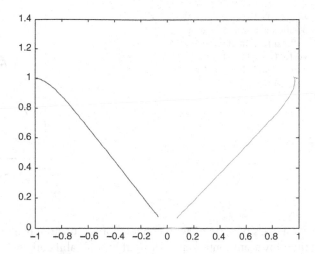

Example 2.1 We illustrate by a numerical example. The task is to get a unicycle to drive to a beacon placed at the origin from any starting point. We start from the feedback linearized model (2.9). We can drive the point p to the origin by the control law $u = -p$, i.e., $u = -z - \varepsilon r$. From (2.8)

$$v = -\langle r, z \rangle - \varepsilon, \quad \omega = -\varepsilon^{-1} \langle s, z \rangle.$$

Now our control laws are

$$v = -x \cos(\theta) - y \sin(\theta) - \varepsilon, \quad \omega = \varepsilon^{-1} (x \sin(\theta) - y \cos(\theta)).$$

Figure 2.12 shows simulation results. Because the robot is not initially headed toward the origin, there are initial turns, followed by straight line segments. The robot does not meet the origin. To get it to end up closer to the origin, one would have to make ε smaller. Again, this makes ω have larger values, as can be seen from the formula

$$\omega(t) = \varepsilon^{-1} \left[x(t) \sin(\theta(t)) - y(t) \cos(\theta(t)) \right].$$

For $t = 0$, $x(0) = 1$, $y(0) = 1$, $\theta(0) = 0$, we have $\omega(0) = \varepsilon^{-1}$. $\quad\triangle$

2.3 Stabilizing the Unicycle to the Origin

Consider the problem of stabilizing the unicycle to the origin. This problem is of little practical interest, but it illustrates some of the challenges in dealing with systems with nonholonomic constraints. A consequence of a celebrated result by Brockett [5] is that

Fig. 2.13 The robot is
positioned at z and heading
in the direction of r. The
vector $0 - z$ from the robot
to the beacon has coordinates
(x_b, y_b) in the frame $\{r, s\}$

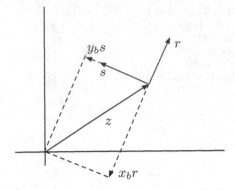

for a class of systems with nonholonomic constraints there do not exist continuous
time-invariant control laws for equilibrium stabilization. The unicycle and the bicycle
are examples of such systems.

While it is impossible to stabilize the unicycle to the origin by means of a continuous time-invariant control law, it *is* possible to do so by means of a continuous (in
fact, smooth) *time-varying* control law. We start with the unicycle model

$$\dot{x} = v\cos(\theta)$$
$$\dot{y} = v\sin(\theta)$$
$$\dot{\theta} = \omega.$$

Here it is convenient to decomplexify the quantities z and r defined earlier and regard
them as real vectors. Thus $z = (x, y)$ and $r = (\cos(\theta), \sin(\theta))$.

Suppose the unicycle mounts a camera pointing in the direction of the body frame
vector r. Suppose also that there is a beacon at the origin, and that the camera is able
to measure the coordinates (x_b, y_b) of the beacon in the body frame $\mathcal{B}$. To derive
(x_b, y_b), consider Fig. 2.13. The displacement of the beacon relative to the unicycle
is the vector $-z$. The scalars x_b and y_b are the projections of this vector onto the body
frame axes $\{r, s\}$. Therefore

$$x_b = \langle -z, r \rangle = \begin{bmatrix} -x & -y \end{bmatrix} \begin{bmatrix} \cos(\theta) \\ \sin(\theta) \end{bmatrix} = -x\cos(\theta) - y\sin(\theta)$$

$$y_b = \langle -z, s \rangle = \begin{bmatrix} -x & -y \end{bmatrix} \begin{bmatrix} -\sin(\theta) \\ \cos(\theta) \end{bmatrix} = x\sin(\theta) - y\cos(\theta).$$

The controller equations are taken to be

$$v = kx_b$$
$$= -k[x\cos(\theta) + y\sin(\theta)]$$
$$w(t) = \cos(t),$$

where k is a small positive gain. Note that only the measurement x_b is needed, not y_b. The usefulness of the periodic w will be revealed soon.

The position dynamics of the closed-loop unicycle are given by

$$\dot{z} = -kz^T rr$$
$$= -krr^T z.$$

Define the 2×2 matrix $M = rr^T$. Then

$$\dot{z} = -kMz.$$

Now look at M:

$$r = (\cos(\theta), \sin(\theta))$$
$$M = rr^T$$
$$-\begin{bmatrix} \cos(\theta) \\ \sin(\theta) \end{bmatrix} \begin{bmatrix} \cos(\theta) & \sin(\theta) \end{bmatrix}$$
$$= \begin{bmatrix} \cos^2(\theta) & \cos(\theta)\sin(\theta) \\ \cos(\theta)\sin(\theta) & \sin^2(\theta) \end{bmatrix}.$$

Since $\theta(t) = \theta(t_0) + \sin(t)$, so $M(t)$ is a 2π-periodic function of t.

Whether or not the unicycle converges to the origin reduces to studying a periodically time-varying linear system. This is a *post facto* motivation for the control law $w(t) = \cos(t)$.

Look at the function $\cos^2(\theta(t))$. Its average value over one period is

$$\frac{1}{2\pi} \int_0^{2\pi} \cos^2(\theta(t))dt.$$

Likewise, the average of $M(t)$ is

$$\overline{M} = \begin{bmatrix} m_1 & m_2 \\ m_2 & m_3 \end{bmatrix}$$
$$= \begin{bmatrix} \frac{1}{2\pi}\int_0^{2\pi}\cos^2(\theta(t))dt & \frac{1}{2\pi}\int_0^{2\pi}\cos(\theta(t))\sin(\theta(t))dt \\ \frac{1}{2\pi}\int_0^{2\pi}\cos(\theta(t))\sin(\theta(t))dt & \frac{1}{2\pi}\int_0^{2\pi}\sin^2(\theta(t))dt \end{bmatrix}.$$

Lemma 2.2 $\overline{M}$ *is positive definite.*

Proof A symmetric matrix is positive definite if and only if its principle minors are positive. Since $m_1 > 0$, we just have to show $\det(\overline{M}) > 0$, i.e., $m_1 m_3 > m_2^2$.

We have already used the letters x, y, but for this proof alone let us use $x(t) = \cos(\theta(t))$ and $y(t) = \sin(\theta(t))$. Also, we shall temporarily use the inner product

$$\langle x, y \rangle = \frac{1}{2\pi} \int_0^{2\pi} x(t)y(t)dt.$$

Then $m_2^2 < m_1 m_3$ is equivalent to

$$\langle x, y \rangle^2 < \langle x, x \rangle \langle y, y \rangle.$$

The Cauchy–Schwarz inequality gives

$$\langle x, y \rangle^2 \leq \langle x, x \rangle \langle y, y \rangle$$

with equality holding if and only if x is a scalar multiple of y or vice versa. Neither is the case here—since $\theta(t)$ is time-varying, we cannot have $\cos(\theta(t)) = c \sin(\theta(t))$. $\qquad\qquad\square$

With the periodically time-varying (PTV) linear system

$$\dot{z}(t) = -kM(t)z(t)$$

we associate the linear time-invariant (LTI) **averaged system**

$$\dot{z}(t) = -k\overline{M}z(t).$$

Convergence in the LTI system is immediate since $\overline{M}$ is positive definite. We have to show that this implies convergence in the PTV system for small enough k. This uses **averaging theory**.

We begin by reviewing the general linear time-varying system

$$\dot{x}(t) = A(t)x(t). \tag{2.10}$$

The transition matrix of (2.10) is the matrix that maps the state at one time, say t_0, to the state at another time, say t:

$$x(t) = \Phi(t, t_0)x(t_0).$$

In general, there is no closed-form expression for $\Phi(t, t_0)$ in terms of $A(t)$ except in some special cases.

1. As you well know, if $A(t) = A$, a constant matrix, then

$$\Phi(t, t_0) = e^{A(t-t_0)}.$$

2. If $A(t)$ is a scalar (1×1 matrix), then

$$\Phi(t, t_0) = e^{\int_{t_0}^{t} A(\tau)d\tau}.$$

3. If, for every value of t_1 and t_2, $A(t_2)$ and $\int_{t_1}^{t_2} A(\tau)d\tau$ commute, then

$$\Phi(t, t_0) = e^{\int_{t_0}^{t} A(\tau)d\tau}.$$

Theorem 2.1 *Let $A(t)$ be periodic of period T. Suppose that*

$$\bar{A} = \frac{1}{T} \int_0^T A(\sigma)d\sigma$$

has all its eigenvalues in the half plane $\Re(s) < 0$. Then there exists $\varepsilon_0 > 0$ such that the origin of

$$\dot{x}(t) = \varepsilon A(t)x(t)$$

is exponentially stable for all $0 < \varepsilon < \varepsilon_0$.

You can find the proof in [19].

2.3.1 Summary

The simple controller

$$v = -kz^T r, \quad \omega(t) = \cos(t)$$

makes the position of the unicycle converge to the origin for any initial condition. But the proof that it does the job is quite involved. If θ is a known function of t, then the system

$$\dot{x} = v \cos(\theta)$$
$$\dot{y} = v \sin(\theta),$$

with state (x, y) and input v, is linear time-varying.

Chapter 3
The Flocking Problem

3.1 Introduction

In his influential paper [36], Reynolds describes his algorithm to simulate flocks of birds by saying that each simulated bird is implemented as an independent actor that navigates according to its local perception of the dynamic environment, the laws of simulated physics that rule its motion, and a set of behaviors programmed into it by the "animator." The aggregate motion of the simulated flock is the result of the dense interaction of the relatively simple behaviors of the individual simulated birds.

In this chapter we formulate a version of the flocking problem in which n robots are required to move in the same direction with the same speed. Like Reynolds, we require each robot to be an independent actor that takes decisions based on limited information it can sense about its neighbours. To simplify, for now we think of just robots in the plane modelled as unicycles. Later, in Chap. 6, we consider flying robots.

Under appropriate assumptions, the flocking problem for unicycles has a fascinating connection with the problem of synchronizing coupled oscillators. In this context, the Kuramoto model inspires a control law for flocking. We will establish a connection between this control law and the flocking algorithm investigated by Jadbabaie et al. [16].

3.2 Problem Formulation

We begin with n unicycles moving at unit speed:

$$\dot{z}_i = e^{j\theta_i}$$
$$\dot{\theta}_i = \omega_i.$$

© The Author(s) 2016

B.A. Francis and M. Maggiore, *Flocking and Rendezvous in Distributed Robotics*, SpringerBriefs in Control, Automation and Robotics, DOI 10.1007/978-3-319-24729-8_3

Recall that z_i is the ith robot's position, θ_i its heading angle, and ω_i the steering control. The steering controls are restricted to depend on certain sensed variables as required by the nature of how the cameras function. We denote by $r_i = e^{j\theta_i}$ the unit vector tangent to the trajectory. This is the velocity vector of unicycle i. Synchronizing the velocity vectors of the unicycles corresponds to synchronizing the heading angles θ_i.

Let z denote the vector whose components are the unicycle positions:

$$z = (z_1, \ldots, z_n).$$

The **neighbours** of robot i are those robots that are visible by robot i. For example, if robot i carries an omnidirectional camera then its neighbours are those robots whose distances from robot i are not greater than the camera range. The **set of neighbours** of robot i is denoted by $\mathcal{N}_i(z)$ and is a set of indices. We repeat: z_i is the location of robot i (in the global coordinate frame), z is the vector of positions, and $\mathcal{N}_i(z)$ is the set of indices k such that robot k is visible by robot i.[1] Notice that the neighbourhood class $\mathcal{N}_i(z)$ depends on z. The **visibility graph** $\mathcal{G}(z)$ is defined to have n nodes and an edge between two nodes if they are neighbours. If the robot carries an omnidirectional camera, $\mathcal{G}(z)$ is an undirected graph: if robot i can see robot k, then robot k can see robot i. If the camera is not omnidirectional, $\mathcal{G}(z)$ is a directed graph.

Next, we need to define what steering control laws ω_i are **admissible**. This is the case if they are locally Lipschitz and they depend only on sensed data from the onboard cameras. Let robot k be a neighbour of robot i. We assume robot i can see the position and heading of robot k in its local frame; that is, vectors $z_k - z_i$ and r_k in the local frame of robot i. To repeat, if robot k is a neighbour of robot i, then robot i has the following sensed data:

$$\langle z_k - z_i, r_i \rangle, \quad \langle z_k - z_i, s_i \rangle, \quad \langle r_k, r_i \rangle, \quad \langle r_k, s_i \rangle,$$

or equivalently,

$$\langle z_k - z_i, r_i \rangle, \quad \langle z_k - z_i, s_i \rangle, \quad \theta_k - \theta_i.$$

Recall that s is the counterclockwise rotation of r by $\pi/2$: $s = jr$. Figure 3.1 shows the definition of the sensed data for two robots.

The **flocking problem** is to find, if they exist, admissible steering controls ω_i so that there exists $\varepsilon > 0$ such that for all initial conditions satisfying

(i) $(\forall i, k = 1, \ldots, n) \; |\theta_i(0) - \theta_k(0)| < \varepsilon$,
(ii) $\mathcal{G}(z(0))$ is connected,

then

---

[1] In a given dynamical system model, the point locations will depend on time and we shall write, for example, $z_i(t)$. In this way the neighbour set will depend on time: $\mathcal{N}_i(z(t))$. But it would be incorrect to write the time-dependence as $\mathcal{N}_i(t)$, because knowing just t, and not, say, initial locations, is not enough to know the neighbour set in general.

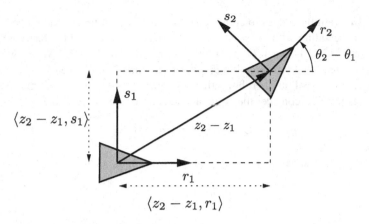

Fig. 3.1 Sensed data for two unicycles

$$(\exists \theta_{ss}) \lim_{t \to \infty} \theta(t) = \theta_{ss}\mathbf{1}.$$

Thus we require that if initially the robots' heading angles are sufficiently close to each other and the visibility graph is connected, then the heading angles will synchronize. Notice that this is a mathematical problem and that the following real issues are not addressed: robots are not of zero dimensions and therefore collisions are a real issue; real robots cannot be programmed all to go at the same speed and therefore a control system would be required to keep the robots in a pack.

Despite the apparent simplicity of the formulation above, the flocking problem remains unsolved. The challenge is to design admissible control laws that preserve the connectivity of the visibility graph while achieving synchronization of the heading angles. If we make the unrealistic assumption that the visibility graph is fixed, i.e., the neighbour sets $\mathcal{N}_i$ are constant, then the flocking problem admits a simple solution, which we discuss next.

3.3 Flocking with Fixed Neighbours

If the visibility graph $\mathcal{G}$ is fixed (i.e., the neighbour sets $\mathcal{N}_i$ are constant) and connected, then we no longer have to worry that the robots might come in and out of each other's field of view. In this case we may ignore the robots' positions and consider just their heading angle dynamics

$$\dot{\theta}_i = \omega_i, \quad i = 1, \ldots, n. \tag{3.1}$$

In this context, we seek controls ω_i that rely only on the sensed data $\{\theta_i - \theta_k : k \in \mathcal{N}_i\}$, and asymptotically stabilize the **flocking manifold**

$$\mathcal{F} = \{(\theta_1, \ldots, \theta_n) : \theta_1 = \cdots = \theta_n\}. \tag{3.2}$$

We will also want the angles to converge to a constant, not just to each other.

Before proceeding with the solution of this problem, we need to clarify what the state space of (3.1) is. Each θ_i is an angle measured modulo 2π. When we write $\theta_i = \theta_k$, we mean that θ_i is equal to θ_k plus some integer multiple of 2π. An angle, therefore, is not a real number, but rather an *equivalence class* of real numbers. To make things precise, consider the equivalence relation on the real line

$$\theta_1 \sim \theta_2 \iff (\exists l \in \mathbb{Z})\, \theta_1 - \theta_2 = 2\pi l.$$

For $\theta \in \mathbb{R}$, denote by $[\theta]$ the equivalence class of θ, defined as

$$[\theta] := \{\theta + 2\pi l : l \in \mathbb{Z}\}.$$

Finally, let $(\mathbb{R}/\sim)$ denote the set of all equivalence classes just defined. Then each angle lives in $(\mathbb{R}/\sim)$, and the state space of (3.1) is the n-fold product $(\mathbb{R}/\sim) \times \cdots \times (\mathbb{R}/\sim)$.

How can we visualize the set $(\mathbb{R}/\sim) \times \cdots \times (\mathbb{R}/\sim)$? First, it is not a vector space, for the sum of two angles is well-defined but scalar multiplication is not (try to multiply $[\pi]$ by $1/2$; do you get a unique angle in $(\mathbb{R}/\sim)$?). A useful way of visualizing $(\mathbb{R}/\sim)$ is to identify each element $[\theta] \in (\mathbb{R}/\sim)$ with the point $e^{j\theta}$ on the unit circle in the complex plane, which we denote by $\mathbb{S}^1$. For an exercise, convince yourself that this identification works. Prove that the map $(\mathbb{R}/\sim) \to \mathbb{S}^1$, $[\theta] \mapsto e^{j\theta}$ is a bijection.[2] From now on we will drop the notation $[\theta]$ and use θ to denote an angle in $(\mathbb{R}/\sim)$. We will also identify angles with points on the unit circle.

To summarize, the state space of (3.1) is the n-fold product $\mathbb{S}^1 \times \cdots \times \mathbb{S}^1$, called the n-torus, and the flocking problem with fixed visibility graph corresponds to the synchronization of points on the unit circle. Since the n-torus is not a vector space, it is not possible to solve the flocking problem with a linear control law. More concretely, for a control law to be well-defined on the n-torus, it must be a 2π-periodic function of the angles $(\theta_1, \ldots, \theta_n)$, and therefore nonlinear.

In 1975 the Japanese researcher Yoshiki Kuramoto proposed a simple model of synchronization of coupled oscillators. Kuramoto represented the limit cycle behaviour of each oscillator by the motion of a point θ_i on the unit circle $\mathbb{S}^1$, and formulated the following coupled differential equation representing the interconnection of the oscillators:

$$\dot{\theta}_i = \mu_i - \frac{K}{n} \sum_{k=1}^{n} \sin(\theta_i - \theta_k), i = 1, \ldots, n. \tag{3.3}$$

The scalar $\mu_i \geq 0$ represents the ith oscillator's natural frequency and $K > 0$ determines the strength of the coupling among oscillators. This equation is known as the **Kuramoto model** of coupled oscillators. In a moment we will derive the Kuramoto model, but first we remark on its qualitative properties.

---

[2]In fact, the set $(\mathbb{R}/\sim)$ can be given a differentiable structure turning it into a smooth manifold, and such that the map $(\mathbb{R}/\sim) \to \mathbb{S}^1$ defined here is a diffeomorphism, not just a bijection.

When the natural frequencies μ_i are positive, Kuramoto observed that as K is increased beyond a minimum threshold K_{min} (the so-called critical coupling strength), some of the points θ_i begin to rotate around the unit circle in a cohesive group, although not synchronized. The larger the ratio K / K_{min}, the greater the portion of oscillators moving cohesively and the smaller the distance between the θ_i's. In the limit as $K \to \infty$, all θ_i's become synchronized. Kuramoto's results stimulated a vigorous research activity in the physics and dynamical systems communities. Some of that research is summarized in the paper by Strogatz [41]. Recently, Dörfler and Bullo [10] made a breakthrough by precisely characterizing, among other things, the critical coupling strength K_{min}, the presence of a locally exponentially stable set in the state space of (3.3), the size of the domain of attraction of this set, and the bound on the distances between the θ_i's on this set.

When the natural frequencies μ_i are all zero, the qualitative properties of model (3.3) are much simpler to analyze. As a matter of fact, we will see that for any $K > 0$ the manifold[3] $\{(\theta_1, \ldots, \theta_n) : \theta_1 = \cdots = \theta_n\}$ is locally asymptotically stable. In particular, when the initial conditions $\theta_i(0)$ are sufficiently close to each other, all phases θ_i converge to a common constant θ_{ss}. This result is precisely what we need to solve the flocking problem. Before stating a formal result, we derive the Kuramoto model.

3.3.1 Derivation of the Kuramoto Model

Consider a collection of points $p_1(t), \ldots, p_n(t)$ moving on the unit circle, with $p_i(t) = e^{j\theta_i(t)}$. Differentiate with respect to time: $\dot{p}_i = e^{j\theta_i} j\dot{\theta}_i$. Now let $v_i = \dot{\theta}_i$ and substitute into the preceding equation:

$$\dot{p}_i = v_i j p_i. \tag{3.4}$$

Since the circle has unit radius, the scalar v_i represents the linear speed of the point p_i. This speed could be positive or negative. Notice that if we view p_i as a vector from the origin and view multiplication by j as rotation by $\pi/2$, then jp_i can be viewed as tangent to the circle at the point p_i—see the picture on the left in Fig. 3.2.

Now we propose a feedback law for v_i in Eq. (3.4); see the picture on the right in Fig. 3.2. First we suppose that when there are no interactions, point p_i has a nominal speed μ_i. To model the interaction between p_i and p_k, we add a term proportional to the projection of p_k onto the line tangent to the circle at p_i, that is, $\langle p_k, jp_i \rangle$. We then sum all these contributions to arrive at the control law

$$v_i = \mu_i + \frac{K}{n} \sum_{k=1}^{n} \langle p_k, jp_i \rangle.$$

---

[3]Recall that we called this set the flocking manifold $\mathcal{F}$ of the model (3.1).

Fig. 3.2 *Left* The vectors
p_i and jp_i. *Right* The
contribution to the linear
speed v_i of the interaction
between points i and k

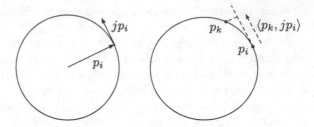

The coefficient K/n determines the interaction strength between points.

Now we need to express v_i in terms of the angles θ_k rather than the points p_k. We observe that

$$\langle p_k, jp_i \rangle = \text{Re}\,(\bar{p}_k jp_i)$$
$$= \text{Re}\left(e^{-j\theta_k} je^{j\theta_i}\right)$$
$$= -\sin(\theta_k - \theta_i).$$

Therefore,

$$\dot{\theta}_i = v_i = \mu_i - \frac{K}{n}\sum_{k=1}^{n}\sin(\theta_i - \theta_k).$$

This is the Kuramoto model in (3.3).

3.3.2 Solution of the Flocking Problem

We now return to the flocking problem for system (3.1). Kuramoto's model (3.3) with natural frequencies $\mu_i = 0$ suggests the following choice[4] of control laws:

$$\omega_i = -\sum_{k=1}^{n} a_{ik}\sin(\theta_i - \theta_k), \quad a_{ik} > 0.$$

These controls require each robot to see all other robots (i.e., the visibility graph $\mathcal{G}$ is complete). It turns out that we may simply restrict the sum to the index set of the neighbours of robot i:

---

[4]For maximum generality, we replace the coefficient K/n in the Kuramoto model by symmetric gains $a_{ik} = a_{ki} > 0$.

$$\omega_i = -\sum_{k\in\mathcal{N}_i} a_{ik}\sin(\theta_i - \theta_k), \quad a_{ik} = a_{ki} > 0. \tag{3.5}$$

Theorem 3.1 *If the visibility graph is fixed, undirected, and connected, then for any $a_{ik} = a_{ki} > 0$, $i, k = 1, \ldots, n$, the control law (3.5) makes the flocking manifold $\mathcal{F}$ in (3.2) locally asymptotically stable and solves the flocking problem.*

Proof Sketch Suppose for simplicity that the visibility graph is complete and $a_{ik} = 1$, so that the closed-loop system reads as

$$\dot{\theta}_i = -\sum_{k=1}^{n} \sin(\theta_i - \theta_k), \quad i = 1, \ldots, n. \tag{3.6}$$

This equation has the form $\dot{\theta} = g(\theta)$, where the $g(\theta)$ is the gradient of a positive definite function. Indeed, let $re^{j\psi}$ denote the average of the points $e^{j\theta_1}, \ldots, e^{j\theta_n}$. Of course, r and ψ are functions of θ, and so we have

$$r(\theta)e^{j\psi(\theta)} = \frac{1}{n}\left(e^{j\theta_1} + \cdots + e^{j\theta_n}\right)$$

and therefore

$$r(\theta) = \frac{1}{n}\left|e^{j\theta_1} + \cdots + e^{j\theta_n}\right|.$$

The average of n points on the unit circle lives inside the unit disc, and therefore $r(\theta)$ is a real number between 0 and 1. It equals 1 if and only if the n points are equal, that is, the n angles are equal, and this is the case when the angles are on the flocking manifold $\mathcal{F}$.

Define the function

$$V(\theta) = \frac{n^2}{2}r(\theta)^2$$

$$= \frac{1}{2}\left|e^{j\theta_1} + \cdots + e^{j\theta_n}\right|^2$$

$$= \frac{1}{2}\left(e^{j\theta_1} + \cdots + e^{j\theta_n}\right)\left(e^{-j\theta_1} + \cdots + e^{-j\theta_n}\right).$$

Thus

$$\frac{\partial V(\theta)}{\partial\theta_i} = \sin(\theta_1 - \theta_i) + \cdots + \sin(\theta_n - \theta_i)$$

and therefore (3.6) can be written $\dot{\theta} = \partial V(\theta)/\partial\theta$. This is a gradient equation. If $\theta(0)$ is chosen so that all the phases are close enough together, then $r(\theta(0))$ will be close to 1, and therefore θ will move in a direction to increase $V(\theta)$, that is, increase $r(\theta)$, until in the limit $r(\theta) = 1$ and the phases are synchronized.

In the general case of $a_{ik} \neq 1$ and connected but not complete graph, the function V should be replaced by

$$V(\theta) = -\sum_{i=1}^{n} \sum_{k \in \mathcal{N}_i, k < i} a_{ik}(1 - \cos(\theta_i - \theta_k)).$$

We will see in Chap. 6 that $-V(\theta)$ can be regarded as the potential of a collection of springs with constants a_{ik} that connect point-masses on the unit circle.

With the above definition of V the closed-loop system with controls (3.5) can still be written as $\dot{\theta} = \partial V(\theta)/\partial \theta$, and it is still true that solutions in a neighbourhood of the global maximum of V converge to the global maximum. It can be shown that if the graph is connected, the global maximum of V is attained on the flocking manifold $\mathcal{F}$. The details of this argument are presented later in the proof of a flocking theorem for flying robots (Theorem 6.1).   □

3.4 The Control Law of Jadbabaie, Lin, and Morse

The paper by Jadbabaie, Lin, and Morse [16] is on the flocking problem. Since it has been referenced by many other papers, it is instructive to review it from the viewpoint of this chapter.

Although it is not stated in these terms, the paper studies a system of unicycles moving at constant speed in the plane. It is assumed that if unicycle i can see unicycle j, then j can see i. That is, the visibility graph is undirected at all times. The paper studies a single control strategy in discrete time, namely, at time $t + 1$ (t an integer) unicycle i changes its heading to the average heading at time t of itself and its neighbours:

$$\theta_i(t + 1) = \frac{1}{1 + n_i(t)} \left(\theta_i(t) + \sum_{k \in \mathcal{N}_i(t)} \theta_k(t) \right). \tag{3.7}$$

Here $n_i(t)$ is the number of neighbours of robot i at time t. It is proved that all the heading angles converge to a common value provided $\mathcal{G}(t)$ has a connectedness property over time, namely, there exists $T > 0$ such that the union graph $\bigcup_{t_0 \leq t \leq t_0 + T} \mathcal{G}(t)$ is connected for all t_0. Here, the union of graphs with the same node set is obtained by taking the union of the edges. Unfortunately, the condition on $\mathcal{G}(t)$ is not checkable—it would require an infinite time simulation. The proof uses a theorem of Wolfowitz on ergodicity (1963).

In the theorems in [16] the vector $\theta(0)$ of initial heading angles is fixed in time. This allows the visibility graph to be a function of t alone, and not both t and $\theta(0)$. In actuality, as we remarked at the beginning of this chapter, for a sensible model of limited visibility, the graph is state dependent, $\mathcal{G} = \mathcal{G}(z)$. If the control strategy is given and if the state $z(t)$ evolves uniquely from $z(0)$, the visibility graph is a function of time, $\mathcal{G}(z(t))$.

Is the control law of [16] admissible? If we ignore the problem with the time-varying visibility graph, it is admissible in the following sense. In continuous time, consider the controls

$$\omega_i = -\frac{1}{n_i(t) + 1} \sum_{k \in \mathcal{N}_i(t)} (\theta_i - \theta_k). \tag{3.8}$$

These are admissible because they rely on relative angles of neighbouring robots. The forward-Euler discretization with unit sampling interval of the closed-loop system is

$$\theta_i(t + 1) = \theta_i(t) - \frac{1}{n_i(t) + 1} \sum_{k \in \mathcal{N}_i(t)} (\theta_i(t) - \theta_k(t)),$$

which coincides with (3.7). Thus the update (3.7) results from a forward-Euler discretization of an admissible control law.

The update law (3.7) is not 2π-periodic. To illustrate the relevance of this property, consider three robots ($n = 3$) and assume that 1 can see 2, 2 can see 1 and 3, and 3 can see 2. Initialize the heading angles as follows: $\theta_1(0) = 0$, $\theta_2(0) = 2\pi$, $\theta_3(0) = 4\pi$. Thus the initial headings are identical modulo 2π. At the next time step we have

$$\theta_1(1) = \frac{1}{2}(0 + 2\pi) = \pi$$

$$\theta_2(1) = \frac{1}{3}(2\pi + 0 + 4\pi) = 2\pi$$

$$\theta_3(1) = \frac{1}{2}(4\pi + 2\pi) = 3\pi.$$

Thus the heading angles move away from the flocking manifold. The flocking manifold is unstable. More precisely, the problem with the control law (3.7) is that it treats the angles θ_i as real numbers, rather than as elements of $\mathbb{S}^1$. The control law (3.5), on the other hand, is 2π-periodic and it respects the geometry of the state space.

As a final remark, notice the relationship between the Kuramoto-inspired control law (3.5) and the control law (3.8). If the visibility graph is constant and undirected, and we set $a_{ik} = 1/(n_i + 1)$, then (3.8) is the linearization of (3.5) at any point of the flocking manifold.

Chapter 4
The Rendezvous Problem: Fixed Neighbours

4.1 Introduction

A dictionary definition of *rendezvous* is, for example, a meeting of people at a pre-designated place and time. For this to happen, the place must be known by the people beforehand. In the rendezvous problem, however, the place cannot be specified beforehand because a robot, as we define it, has no knowledge of global coordinates.

The **rendezvous problem** for integrator point models can be stated as follows:

Get n identical mobile robots with only onboard sensors to move to a common location using distributed control.

This is also called an agreement or consensus problem. It is a theoretical problem; in practice it could be used to get the robots to gather near to each other. It is also related to the problem of electing a leader. We emphasize that the terminal point is not specified; in general it is unknown until the robots do meet.

All solutions to the rendezvous problem are based on the strategy of pursuit, so we begin this chapter with the simplest such strategy: cyclic pursuit of integrator points. Cyclic pursuit assumes a special kind of visibility graph. Later we generalize cyclic pursuit to arbitrary fixed graphs. The key result is a theorem relating the spectral properties of the Laplacian of the visibility graph to its connectedness.

The rendezvous problem for unicycles is considerably harder than for integrator points. We present a rendezvous strategy for unicycles and highlight its practical limitations. We then return to cyclic pursuit and illustrate some unexpected phenomena arising when there are infinitely many integrator points. We conclude the chapter with a discussion about discrete-time models of robots.

© The Author(s) 2016
B.A. Francis and M. Maggiore, *Flocking and Rendezvous in Distributed Robotics*, SpringerBriefs in Control, Automation and Robotics, DOI 10.1007/978-3-319-24729-8_4

4.2 Cyclic Pursuit

Suppose three dogs are initially placed at the vertices of an equilateral triangle and they chase each other at unit speed. Dog 1 chases dog 2; dog 2 chases dog 3; finally, dog 3 chases dog 1. What curves do the dogs describe? The answer, depicted in Fig. 4.1, is that each dog traces out a logarithmic spiral, and the dogs meet at a common point inside the triangle. The curves of Fig. 4.1 are special instances of **pursuit curves**.

If we let z_i denote the position of dog i in the complex plane, then all pursuit curves traced out by the three dogs in the complex plane arise from solutions of the coupled differential equations

$$\dot{z}_1 = \frac{z_2 - z_1}{|z_2 - z_1|}$$

$$\dot{z}_2 = \frac{z_3 - z_2}{|z_3 - z_2|}$$

$$\dot{z}_3 = \frac{z_1 - z_3}{|z_1 - z_3|}.$$

We may interpret the above equations like this. The dogs are modelled as integrator points $\dot{z}_i = u_i$, and the pursuit strategy of dog i is the control law $u_i = (z_{i+1} - z_i)/|z_{i+1} - z_i|$. This control law is nonlinear. The control law $u_i = (z_{i+1} - z_i)$ is linear and preserves the essential qualitative properties of pursuit curves.

We now generalize the scenario above. Consider a collection of robots modelled as integrator points. The robots are numbered 1 to n. Robot 1 pursues robot 2; it in turn pursues robot 3; and so on; robot $n - 1$ pursues robot n; finally, robot n pursues robot 1. This strategy is called **cyclic pursuit**. Notice that the robots must be identifiable for cyclic pursuit to be implemented. Robot 1 must be able to identify which of the other robots is robot 2, and it must be able to see it at all time.

Fig. 4.1 Pursuit curves

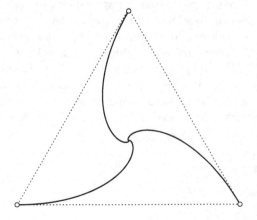

Example 4.1 Consider four robots in cyclic pursuit. Model the robots as points in the complex plane: $z_1, \ldots, z_4 \in \mathbb{C}$. The neighbour sets are $\mathcal{N}_i(z) = \{i + 1\}$ for $i = 1, 2, 3$ and $\mathcal{N}_4(z) = \{1\}$ and the differential equations are

$$\dot{z}_1 = z_2 - z_1$$
$$\dot{z}_2 = z_3 - z_2$$
$$\dot{z}_3 = z_4 - z_3$$
$$\dot{z}_4 = z_1 - z_4.$$

Define the vector and matrix

$$z = \begin{bmatrix} z_1 \\ z_2 \\ z_3 \\ z_4 \end{bmatrix}, \quad U = \begin{bmatrix} 0 & 1 & 0 & 0 \\ 0 & 0 & 1 & 0 \\ 0 & 0 & 0 & 1 \\ 1 & 0 & 0 & 0 \end{bmatrix}.$$

Then the preceding four equations can be combined into

$$\dot{z} \doteq (U - I)z. \tag{4.1}$$

The four trajectories can be found by familiar eigenvalue analysis. The characteristic polynomial of U is $s^4 - 1$, whose roots, i.e., eigenvalues of U, are $1, j, -1, -j$, the four roots of unity. Thus the eigenvalues of $U - I$ are the four points

$$0, -1 + j, -2, -1 - j$$

as shown in Fig. 4.2. For a given $z(0)$, the solution of (4.1) is the sum of four terms, one for each eigenvalue. The three terms for the left half-plane eigenvalues go to zero as t goes to ∞. The term for the zero eigenvalue is stationary. An eigenvector for

Fig. 4.2 The eigenvalues of $U - I$ are on the circle with centre -1 and radius 1

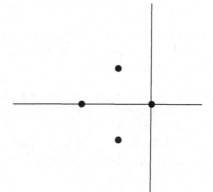

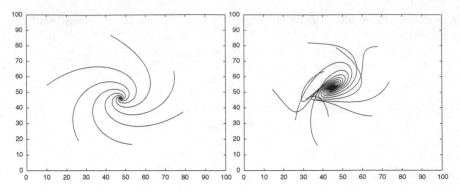

Fig. 4.3 Simulations of cyclic pursuit

this zero eigenvalue is **1**, the vector of all 1's. Thus there exists a complex number c such that $z(t) \to c\mathbf{1}$, or $z_i(t) \to c$ for all i. This proves that the robots rendezvous.

Figure 4.3 shows two simulations (more than 4 robots) to illustrate the behaviour. We observe that the centroid is stationary, each robot converges to the centroid, and the trajectories sometimes intersect, but sometimes do not—it depends on the initial configuration. Furthermore, control is distributed in that the robots have identical local strategies and the sensor requirements are minimal: Only n information-flow links are needed for n robots. Also, there is an **emergent behaviour**—convergence to a common point.   △

More generally, consider finitely many, $n \geq 1$, robots with positions $z_1, \ldots, z_n$ in the complex plane. Let z denote the vector with components z_1 up to z_n. Let U denote the $n \times n$ real matrix with first row $0, 1, 0 \ldots, 0, 0$, second row $0, 0, 1, 0, \ldots, 0$, and so on, until the last row $1, 0, \ldots, 0$. Then the cyclic pursuit model is $\dot{z} = Mz$, $M = U - I$. In components,

$$\dot{z}_i = z_{i+1} - z_i, \quad i = 1, \ldots, n-1$$
$$\dot{z}_n = z_1 - z_n. \tag{4.2}$$

Theorem 4.1 *System (4.2) has the property that the centroid of the robots' positions, $(1/n)\mathbf{1}^T z(t)$, is stationary and for every initial condition, the robots asymptotically rendezvous at this centroid, i.e., $z_i(t) \to (1/n)\mathbf{1}^T z(0)$ for all $i \in \{1, \ldots, n\}$.*

Proof Let us see first that the centroid is stationary. The vector of points is $z(t)$. The centroid of these points equals $(1/n)\mathbf{1}^T z(t)$. The derivative of this equals zero because $\mathbf{1}^T M = 0$.

The behaviour is completely specified by the eigenvalues and eigenvectors of M. Let us first note that U is a companion matrix with characteristic polynomial $s^n - 1$.

The roots of this polynomial, and hence the eigenvalues of U, are the roots of unity, namely, these n points on the unit circle:

$$\text{eigenvalues of } U: \quad e^{j2\pi k/n}, \quad k = 0, 1, \ldots, n-1.$$

By the spectral-mapping theorem, the eigenvalues of $M = U - I$ are these n points shifted left by 1:

$$\text{eigenvalues of } M: \quad e^{j2\pi k/n} - 1, \quad k = 0, 1, \ldots, n-1.$$

One of these is at the origin and all the others are strictly in the left half-plane. Thus every solution of $\dot{z} = Mz$ converges to the eigenspace of the zero eigenvalue. Let $\mathbf{1}$ denote the n-dimensional vector of all 1s. Clearly $U\mathbf{1} = \mathbf{1}$ and hence $M\mathbf{1} = 0$. Thus $\mathbf{1}$ is an eigenvector of M for the zero eigenvalue. In other words, the eigenspace for the zero eigenvalue is one-dimensional and spanned by $\mathbf{1}$. We denote this eigenspace by $\ker M$, where ker denotes kernel, which is also termed the nullspace. It follows from linear system theory that every solution of $\dot{z} = Mz$ converges to $c\mathbf{1}$ for some complex number c dependent on the initial condition. Hence the robots $z_i(t)$ all converge to the same point. Since $(1/n)\mathbf{1}^T z(t)$ is stationary, and since $z(t) \to c\mathbf{1}$, we have

$$(1/n)\mathbf{1}^T z(0) = \lim_{t \to \infty} (1/n)\mathbf{1}^T z(t) = (1/n)\mathbf{1}^T c\mathbf{1} = c.$$

Thus all robots converge to the centroid of their initial positions. □

Notice that the robots are modelled as living in the complex plane, while the aggregate state space is $\mathbb{C}^n$. In $\mathbb{C}$, there are n trajectories, all terminating at the same point. In the state space $\mathbb{C}^n$, there is one trajectory, as shown in Fig. 4.4.

Here is another type of rendezvous. You go to the zoo with a friend. At some time you unfortunately become separated. How can you meet up again? This is a kind of rendezvous problem. If you had pre-arranged that in the eventuality of becoming

Fig. 4.4 The trajectory converges to the subspace $\ker M$

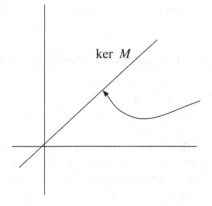

separated you would both go to the entrance (or some other beacon), there would be no problem, so we exclude this situation. Likewise, we exclude the possibility of making an announcement over a PA system. The real problem is to devise identical search procedures for you and your friend to guarantee meeting, preferably in a reasonable time. Another example (from Naomi Leonard): n autonomous underwater vehicles have been exploring the ocean; at the end of a certain time period it is desired that they assemble at a common point so that their power supplies can be recharged. This is a rendezvous problem. The flocking problem with fixed visibility graph of Chap. 3 can be also viewed as a rendezvous problem for points on the unit circle.

We turn to implementation details. Assume each robot has onboard a camera, a computer, and a clock. Also, each robot has a spherical head and the heads are all of a different colour. (Thus one robot can compute the distance to another using a camera by measuring the diameter of the latter's sphere.) Assume there is a supervisor (a human or computer) who can occasionally download instructions to the robots. We think of this as a hybrid control setup: The supervisor at the discrete task level; the robots at the lower level. Prior to the rendezvous task, the supervisor downloads the pursuit instructions to the robots, for example, *red* should pursue *green*. The supervisor also downloads a time value t_0 at which the rendezvous task should begin. Assume that at time $t = 0 < t_0$ the robots are dispersed in some fashion, say, by the supervisor. At $t = t_0$ cyclic pursuit would begin and rendezvous would occur (asymptotically).

In cyclic pursuit, robot i is assumed to see robot $i + 1$ with perfect accuracy and no matter how far apart they are. Here is a more realistic scenario: There are n robots. At time t, robot i can see (or otherwise sense) a subset $\mathcal{N}_i(t)$ of the others. He heads for their centroid (or some other linear combination of their positions). The set $\mathcal{N}_i(t)$ may change with time, as for example if the sensor is a camera with a limited field of view, such as a cone. So the visibility graph is time-varying. Can an emergent collective behaviour be assured? We will study this in Chap. 5.

4.3 General Fixed Neighbours

We now formulate the rendezvous problem and generalize the cyclic pursuit strategy.

4.3.1 Rendezvous Problem for Integrator Points

Let there be n robots:

$$\dot{z}_i = u_i, \quad i = 1, \ldots, n. \tag{4.3}$$

Each robot can see some neighbours using an *onboard* camera. We assume that if robot i can see robot j, then robot i is able to measure the relative displacement $z_j - z_i$. Let $\mathcal{N}_i$ denote the set of sensed neighbours of robot i; $\mathcal{N}_i$ is a subset of the integers from 1 to n, excluding i. Notice that neighbours are not determined by proximity but rather just by definition. Furthermore, the neighbour sets do not change with time—they are fixed.

A control law u_i is **admissible** if it is a locally Lipschitz function of the relative displacements $z_j - z_i, j \in \mathcal{N}_i$. The **rendezvous problem** is to find admissible controls u_i so that for every $z(0)$ there exists a complex number z_{ss} such that the solution $z(t)$ of (4.3) satisfies $\lim_{t \to \infty} z(t) = z_{ss}\mathbf{1}$.

Let us clean up the notation by using vectors and matrices. Let z and u be the vectors with components z_i and u_i. Let n_i be the number of neighbours of robot i and let $y_i \in \mathbb{C}^{n_i}$ be the vector of relative displacements sensed by robot i. Thus the components of y_i are the quantities $z_j - z_i, j \in \mathcal{N}_i$. We may write $y_i = C_i z$, where C_i is a matrix $n_i \times n$. If $n_i = 0$ then we set C_i to be the zero $1 \times n$ matrix. Admissibility of the control law u_i corresponds to requiring u_i to be a function of y_i. We consider linear control laws, $u_i = F_i y_i$, where F_i is a matrix of dimension $1 \times n_i$. Now define

$$y = \begin{bmatrix} y_1 \\ \vdots \\ y_n \end{bmatrix}, \quad C = \begin{bmatrix} C_1 \\ \vdots \\ C_n \end{bmatrix}, \quad F = \begin{bmatrix} F_1 & 0 & \cdots & 0 \\ 0 & F_2 & \cdots & 0 \\ \vdots & \vdots & \vdots & \vdots \\ 0 & 0 & \cdots & F_n \end{bmatrix}.$$

Then $y = Cz$ and the combined control law is $u = Fy$.

To recap, the equations are

$$\dot{z} = u, \quad y = Cz, \quad u = Fy.$$

The closed-loop system is simply $\dot{z} = FCz$. These equations look very simple, but the structure of the problem is embedded in the special forms of C and F. Each row of C is either all 0's (if the corresponding robot has no neighbours) or has only two nonzero elements, $+1$ and -1, and F is block diagonal. These reflect the facts that the sensors are onboard and control is decentralized.

Now we turn to graphical matters. Associated with the neighbour structure is a **visibility graph** $\mathcal{G}$. As in Chap. 3, there is one node for each robot, and an edge from i to j if i can see j, which we interpret as saying $j \in \mathcal{N}_i$. The **adjacency matrix** of the graph is the $n \times n$ matrix A defined by saying

$a_{ij} = 1$ if there is an edge from node i to node j, and
$a_{ij} = 0$ otherwise.

The **out-degree** of a node is the number of edges leaving it. The **degree matrix** of the graph is the diagonal matrix D whose ith diagonal element is the out-degree of node i. Finally, the **Laplacian matrix** of the graph is $L := D - A$. If the visibility graph is undirected, then the Laplacian matrix is symmetric.

Example 4.2 Consider four robots in cyclic pursuit. The visibility graph is shown in Fig. 4.5. The neighbour sets are

$$\mathcal{N}_1 = \{2\}, \quad \mathcal{N}_2 = \{3\}, \quad \mathcal{N}_3 = \{4\}, \quad \mathcal{N}_4 = \{1\}.$$

Fig. 4.5 Visibility graph for
four robots in cyclic pursuit

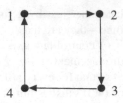

The sensed outputs are

$$y_1 = z_2 - z_1$$
$$y_2 = z_3 - z_2$$
$$y_3 = z_4 - z_3$$
$$y_4 = z_1 - z_4.$$

Pick unity control gains, $F_i = 1$. Then the closed-loop system is given by

$$\dot{z} = FCz = \begin{bmatrix} -1 & 1 & 0 & 0 \\ 0 & -1 & 1 & 0 \\ 0 & 0 & -1 & 1 \\ 1 & 0 & 0 & -1 \end{bmatrix} z.$$

This is the cyclic pursuit strategy (4.2). The Laplacian matrix of the graph in Fig. 4.5 is

$$L = \begin{bmatrix} 1 & -1 & 0 & 0 \\ 0 & 1 & -1 & 0 \\ 0 & 0 & 1 & -1 \\ -1 & 0 & 0 & 1 \end{bmatrix}.$$

Thus we have the identity $FC = -L$ and the dynamical system is therefore

$$\dot{z} = -Lz. \tag{4.4}$$

This is an important equation, showing that robot dynamics is related to the system's graph through the graph's Laplacian.   △

Example 4.3 In our setup, everything is uniquely determined by specifying just the number of robots and the neighbour sets. Consider $n = 4$ robots with neighbour sets

$$\mathcal{N}_1 = \{2\}, \quad \mathcal{N}_2 = \{3\}, \quad \mathcal{N}_3 = \{2, 4\}, \quad \mathcal{N}_4 = \{3\}.$$

Fig. 4.6 A visibility graph

Then the visibility graph is shown in Fig. 4.6. The sensed outputs are

$$y_1 = z_2 - z_1$$
$$y_2 = z_3 - z_2$$
$$y_3 = \begin{bmatrix} z_2 - z_3 \\ z_4 - z_3 \end{bmatrix}$$
$$y_4 = z_3 - z_4.$$

and therefore

$$C = \left[\begin{array}{cccc} -1 & 1 & 0 & 0 \\ 0 & -1 & 1 & 0 \\ \hline 0 & 1 & -1 & 0 \\ \hline 0 & 0 & -1 & 1 \\ \hline 0 & 0 & 1 & -1 \end{array}\right].$$

In general, the matrices C_i are not really unique, but we can make them so if we specify the order in which we will list the relative displacements

$$z_j - z_i, \quad j \in \mathcal{N}_i, \quad i = 1, \ldots, n$$

in the vector y_i. The natural order is for increasing j.

The adjacency and degree matrices of the graph in Fig. 4.6 are

$$A = \begin{bmatrix} 0 & 1 & 0 & 0 \\ 0 & 0 & 1 & 0 \\ 0 & 1 & 0 & 1 \\ 0 & 0 & 1 & 0 \end{bmatrix}, \quad D = \begin{bmatrix} 1 & 0 & 0 & 0 \\ 0 & 1 & 0 & 0 \\ 0 & 0 & 2 & 0 \\ 0 & 0 & 0 & 1 \end{bmatrix},$$

and therefore the Laplacian is

$$L = D - A = \begin{bmatrix} 1 & -1 & 0 & 0 \\ 0 & 1 & -1 & 0 \\ 0 & -1 & 2 & -1 \\ 0 & 0 & -1 & 1 \end{bmatrix}.$$

Note the important property of this decentralized control structure: If $y_1 = 0$, that is, if z_1 is collocated with the only agent it senses, then $u_1 = 0$, so $\dot{z}_1 = 0$ and robot 1 does not move. The matrix F of local controllers has the form

Fig. 4.7 Block diagram

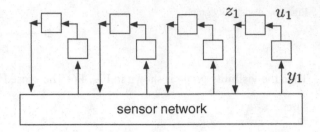

$$F = \begin{bmatrix} F_1 & 0 & 0 & 0 \\ 0 & F_2 & 0 & 0 \\ 0 & 0 & F_3 & 0 \\ 0 & 0 & 0 & F_4 \end{bmatrix}.$$

The components F_1, F_2, F_4 are 1×1 and F_3 is 1×2. The decentralized structure is shown by the block diagram in Fig. 4.7. This is a bit misleading because the concept of a "sensor network" does not bring out the point that the sensors are all local. It is a network of mobile sensors.

Now pick control gains

$$F_1 = 1, \quad F_2 = 1, \quad F_3 = [1 \ 1], \quad F_4 = 1.$$

You can check that the matrix of the closed-loop system is $FC = -L$. Its eigenvalues are $\{0, -1, -1, -3\}$, and the robots rendezvous. $\triangle$

4.3.2 Solution of the Rendezvous Problem

It is a fact that if the rendezvous problem is solvable by some controller whatever, then it is solved by the controller[1]

$$F_i = \mathbf{1}^T. \tag{4.5}$$

To prove this requires a lot of work in graph theory; see [23]. We are going to accept this result. With this F it can be checked that

$$FC = -L.$$

---

[1] The definition of F_i is irrelevant if C_i is the row of zeros, but it is handy to use $F_i = 1$ in this case too.

So now the question becomes, when does L have the property

$$(\forall z(0))(\exists z_{ss}) \lim_{t\to\infty} e^{-tL} z(0) = z_{ss} \mathbf{1}? \tag{4.6}$$

The answer is contained in the following theorem.

Theorem 4.2 *The following conditions are equivalent:*

1. *The rendezvous problem is solvable by some matrices F_i.*
2. *The controller F in (4.5) solves the rendezvous problem.*
3. *The Laplacian L has rank $n - 1$.*
4. *The visibility graph has a node that is reachable from every other node by some directed path.*

Proof   We will prove only equivalence of the second and third conditions. If all the robots are collocated at $t = 0$, they will not move. That is, $C\mathbf{1} = 0$. Therefore $L\mathbf{1} = 0$, from which we conclude that 0 is an eigenvalue of L, that is, the rank of L is less than n. An appeal to Gershgorin's theorem shows that all the other eigenvalues of $-L$ have negative real part. Thus (4.6) will follow if the eigenspace of $-L$ for the zero eigenvalue has dimension not more than 1.   □

Example 4.4   *(Continued from Example 4.3)* As we saw, the Laplacian is

$$L = D - A = \begin{bmatrix} 1 & -1 & 0 & 0 \\ 0 & 1 & -1 & 0 \\ 0 & -1 & 2 & -1 \\ 0 & 0 & -1 & 1 \end{bmatrix}.$$

The rank equals 3 and therefore rendezvous is achievable. The controllers

$$F_1 = 1, \quad F_2 = 1, \quad F_3 = \begin{bmatrix} 1 & 1 \end{bmatrix}, \quad F_4 = 1$$

solve the rendezvous problem.

△

Example 4.5   Let us explore a little further, without proving anything, just doing examples. In the visibility graph, Fig. 4.6, nodes $\{2, 3, 4\}$ have the property that they are reachable from every other node by a directed path. On the other hand, suppose the neighbour sets are

$$\mathcal{N}_1 = \{2\}, \quad \mathcal{N}_2 = \varnothing, \quad \mathcal{N}_3 = \{2, 4\}, \quad \mathcal{N}_4 = \varnothing.$$

Draw the graph. We have

$$L = \begin{bmatrix} 1 & -1 & 0 & 0 \\ 0 & 0 & 0 & 0 \\ 0 & -1 & 2 & -1 \\ 0 & 0 & 0 & 0 \end{bmatrix}.$$

The rank of L equals only 2, which is less than $n - 1 = 3$. The closed-loop system therefore does not enjoy the rendezvous property. The visibility graph shows that there is no node that is reachable from all other nodes by directed paths.

Even though the problem is not solvable, we can continue and find C and F. It will turn out that $FC = -L$ still holds. Based on the neighbour sets, the measured signals are

$$y_1 = z_2 - z_1$$
$$y_2 = 0$$
$$y_3 = \begin{bmatrix} z_2 - z_3 \\ z_4 - z_3 \end{bmatrix}$$
$$y_4 = 0.$$

Thus

$$C = \begin{bmatrix} C_1 \\ C_2 \\ C_3 \\ C_4 \end{bmatrix} = \begin{bmatrix} -1 & 1 & 0 & 0 \\ \hline 0 & 0 & 0 & 0 \\ \hline 0 & 1 & -1 & 0 \\ 0 & 0 & -1 & 1 \\ \hline 0 & 0 & 0 & 0 \end{bmatrix}$$

and

$$F_1 = 1, \quad F_2 = 1, \quad F_3 = \begin{bmatrix} 1 & 1 \end{bmatrix}, \quad F_4 = 1.$$

The equation $FC = -L$ still holds. $\triangle$

As we saw with cyclic pursuit, we have to allow z_{ss} to depend on the initial positions for the goal to be feasible. For example, if all the robots are initially placed at a point $z_i(0) = w$, they will stay there forever. So it would necessarily follow that $z_{ss} = w$.

4.4   Rendezvous of Unicycles

We now turn to rendezvous of unicycles. We present two approaches. In the first approach we feedback linearize the unicycles as in Sect. 2.2 and then use the rendezvous controller for integrator points. In the second approach, we develop a time-varying control law that does not rely on feedback linearization.

Consider n unicycles

$$\dot{z}_i = v_i e^{j\theta_i}$$
$$\dot{\theta}_i = \omega_i. \tag{4.7}$$

Recall that z_i is the position of unicycle i in the complex plane, θ_i is its heading angle, v_i its linear speed, and ω_i its angular speed. As before, we set $r_i := e^{j\theta_i}$ and $s_i := jr_i$. As in Chap. 3, we will say that a control law (v_i, ω_i) is admissible for unicycle i if it is a locally Lipschitz function of the quantities

$$\langle z_k - z_i, r_i \rangle, \quad \langle z_k - z_i, s_i \rangle, \quad \langle r_k, r_i \rangle, \quad \langle r_k, s_i \rangle, \quad k \in \mathcal{N}_i.$$

The rendezvous problem for unicycles is to find admissible controls such that for every $(z(0), \theta(0))$ there exists a complex z_{ss} such that the solution $z(t)$ of (4.7) satisfies $\lim_{t \to \infty} z(t) = z_{ss}\mathbf{1}$. Notice that we do not have any control specification for the heading angles.

Recall from Sect. 2.2 that if we let $p_i := z_i + \varepsilon r_i$ and $u_i := r_i v_i + \varepsilon s_i \omega_i$, then we have

$$\dot{p}_i = u_i.$$

Now we use the main result of the previous section. This requires a redefinition of the sensed outputs y_i in terms of p_i. Accordingly, let y_i denote the vector of differences $p_j - p_i, j \in \mathcal{N}_i$. Then, if the visibility graph satisfies condition 2 of Theorem 4.2, the control law $u_i = \mathbf{1}^T y_i$ makes the positions p_i all converge to the same point. Is this control law admissible? Interestingly, the answer is yes. The control u_i is the sum of terms $p_j - p_i, j \in \mathcal{N}_i$. From u_i we get the physical controls (v_i, ω_i) by solving the equation $u_i = r_i v_i + \varepsilon s_i \omega_i$. Suppose for simplicity that the control law for unicycle 1 is

$$u_1 = p_2 - p_1.$$

Then

$$r_1 v_1 + \varepsilon s_1 \omega_1 = z_2 + \varepsilon r_2 - z_1 - \varepsilon r_1.$$

Taking inner products with r_1 gives

$$v_1 = \langle z_2 - z_1, r_1 \rangle + \varepsilon \langle r_2, r_1 \rangle - \varepsilon$$

and with s_1 gives

$$\omega_1 = \varepsilon^{-1} \langle z_2 - z_1, s_1 \rangle + \langle r_2, s_1 \rangle.$$

Thus the controls are admissible.

However, to get the unicycles almost to rendezvous, we would need ε to be very small. But then the term $\varepsilon^{-1}\langle z_2 - z_1, s_1 \rangle$ in ω_1 would be very large, saturating a real actuator. We conclude that controlling unicycles in this way is feasible but performance is perhaps questionable.

We now turn to the second approach. The following control law does not rely on feedback linearization:

$$v_i = k \sum_{k \in \mathcal{N}_i} \langle z_k - z_i, r_i \rangle$$

$$\omega_i(t) = \cos(t).$$

(4.8)

It has a severe practical limitation: The unicycles keep wiggling even after they have rendezvoused. Nonetheless the following result is interesting.

Theorem 4.3 *Assume that the visibility graph has a node that is reachable from every other node by some directed path. Then there exists $k^\star > 0$ such that for all $k \in (0, k^\star)$, the control law (4.8) solves the rendezvous problem for unicycles.*

Proof Sketch For simplicity we assume that the visibility graph is undirected, so that the Laplacian L is a symmetric matrix. We also decomplexify z_i and view it as a vector in $\mathbb{R}^2$. The positions of the unicycles are modeled by

$$\dot{z}_i = k \sum_{k \in \mathcal{N}_i} \langle z_k - z_i, r_i \rangle r_i, \quad i = 1, \dots, n.$$

Using the identity

$$\langle z, r \rangle r = (r r^T) z,$$

we get

$$\dot{z}_i = k r_i r_i^T \sum_{k \in \mathcal{N}_i} (z_k - z_i), \quad i = 1, \dots, n.$$

(4.9)

Let us express these equations more compactly. To begin with, we may view $\theta_i(t)$ as an exogenous signal because $\dot{\theta}_i = \cos(t)$, so the evolution of θ_i is decoupled from the position dynamics of the unicycles. We suppress t in $\theta_i(t)$ when convenient in what follows. We have the unit direction vector

$$r_i = \begin{bmatrix} \cos(\theta_i) \\ \sin(\theta_i) \end{bmatrix}.$$

Define the matrix

$$M_i := r_i r_i^T = \begin{bmatrix} \cos^2(\theta_i) & \sin(\theta_i)\cos(\theta_i) \\ \sin(\theta_i)\cos(\theta_i) & \sin^2(\theta_i) \end{bmatrix},$$

so that from (4.9)

$$\dot{z}_i = k M_i \sum_{k \in \mathcal{N}_i} (z_k - z_i), \quad i = 1, \dots, n.$$

(4.10)

The position z_i is a function of time t, and because θ_i is also a function of t, so too is M_i a function of t and so we could write $M_i(t)$. Next, let M denote the block diagonal matrix with blocks $\{M_1, \ldots, M_n\}$. Finally, let

$$L_{(2)} = L \otimes I_2,$$

where $\otimes$ denotes the Kronecker product of matrices and I_2 is the 2×2 identity matrix. To recap, element l_{ij} of L is replaced by the block $l_{ij}I_2$. Then (4.10) can be written simply

$$\dot{z} = -kML_{(2)}z,$$

or, with t shown explicitly,

$$\dot{z} = -kM(t)L_{(2)}z. \tag{4.11}$$

This is a linear periodically time-varying system. The analysis from this point onward is similar to what we did in Sect. 2.3 for stabilization of one unicycle to the origin. In Lemma 2.2 we showed that the average of $M(t)$ is positive definite. Denoting by $\bar{M}$ this average, we have the averaged system

$$\dot{z} = -k\bar{M}L_{(2)}z.$$

The matrix $L_{(2)}$ is symmetric. Consider the Lyapunov function candidate

$$V(z) = \frac{1}{2}z^T L_{(2)}z.$$

By Theorem 4.2 L has rank $n - 1$ and its kernel is spanned by $\mathbf{1}$. It follows that $L_{(2)}$ has rank $2(n - 1)$, and its kernel is the image of the $2n \times 2$ matrix $[I_2 \;\cdots\; I_2]^T$. Thus $V(z) = 0$ if and only if $z_1 = \cdots = z_n$, i.e., the robots' positions coincide. The derivative of V along the averaged system is

$$\dot{V} = -kz^T L_{(2)}\bar{M}L_{(2)}z.$$

This is a quadratic function. Since $\bar{M}$ is positive definite, $\dot{V} \leq 0$. Moreover, $\dot{V} = 0$ if and only if $L_{(2)}z = 0$, or $V(z) = 0$. We thus conclude that the set $\{z : V(z) = 0\}$ is globally exponentially stable for the averaged system. Since we are dealing with stability of a set, rather than of an equilibrium, we cannot directly apply Theorem 2.1. The idea is to extract from (4.11) the dynamics transversal to the eigenspace of $L_{(2)}$ associated with the two eigenvalues at zero. Then rendezvous becomes an equilibrium stability problem and we can apply Theorem 2.1. It is possible to show that for sufficiently small k, the set $\{z : V(z) = 0\}$ is globally exponentially stable for the linear time-varying system (4.11), so that the unicycles rendezvous. Moreover $z(t)$ tends to a constant. The details are worked out in [25]. $\qquad\square$

4.5 From Rendezvous to Formation Stabilization

The rendezvous problem is a gateway to more complex control specifications in distributed control. In this section we show that if one can solve the rendezvous problem for integrator points, then one can also make the points converge to any desired formation in a distributed manner. While we focus on integrator points, the discussion can be easily extended to unicycles.[2]

Consider an arbitrary polygon P representing a desired formation for the n robots, for example the triangle displayed in Fig. 4.8. Each vertex of the polygon represents the desired position of a specific robot modulo a translation common to all robots. Label the vertices of P with the indices of the associated robots. To express the polygon as a control specification, place it anywhere in the complex plane and let c_i denote the position of vertex i, as in Fig. 4.8. Finally, let $c = (c_1, \ldots, c_n)$. The vector c is our formation specification.

The **formation stabilization problem** for the integrator points (4.3) is to find admissible controls u_i such that, for every initial condition, there exists $z_{ss} \in \mathbb{C}$ such that the solution $z(t)$ of (4.3) satisfies $z(t) \rightarrow z_{ss}\mathbf{1} + c$. In other words, all robots converge to P modulo translation (but not modulo rotation).

A simple modification of the rendezvous control law $u_i = F_i y_i$, $F_i = \mathbf{1}^T$, solves the formation stabilization problem. Let $d = Lc$, where L is the Laplacian matrix of the visibility graph and define the control laws

$$u_i = \mathbf{1}^T y_i + d_i, \ i = 1, \ldots, n. \tag{4.12}$$

Theorem 4.4 *Suppose the rendezvous problem is solvable for (4.3). Then the above control law solves the formation stabilization problem. In other words, for every initial condition the robots converge to a translated version of the polygon P.*

Proof The closed-loop system is given by

$$\dot{z} = -Lz + d = -L(z - c).$$

Define the error variable $\tilde{z} = z - c$. Then $\dot{\tilde{z}} = -L\tilde{z}$. If the rendezvous problem is solvable, then by Theorem 4.2 for every initial condition there exists a complex number z_{ss} such that $\tilde{z}(t) \rightarrow z_{ss}\mathbf{1}$. Equivalently, $z(t)$ converges to $c + z_{ss}\mathbf{1}$, proving that the robots converge to a translated version of the polygon P. $\square$

Any error in the implementation of the biases d_i will cause the robots to drift. Indeed, suppose that instead of implementing $u_i = F_i y_i + d_i$ we implement $u_i = F_i y_i + d_i + \varepsilon_i$. The closed-loop system in error coordinates becomes

$$\dot{\tilde{z}} = -L\tilde{z} + \varepsilon.$$

---

[2]See [25]. In this case, each unicycle would have to carry a compass or some other device allowing it to measure its own heading angle with respect to a common reference direction.

Fig. 4.8 A polygon P

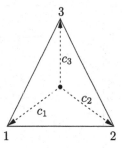

If ε is not in the image of the matrix L, $\tilde{z}(t)$, and therefore $z(t)$, will drift off to infinity.

Example 4.6 Consider six robots in cyclic pursuit. We want to stabilize a triangular formation and a line formation. For the triangle, we choose

$$
c = \begin{bmatrix} 0 \\ 5 - j5\sqrt{3} \\ 10 - j10\sqrt{3} \\ -j10\sqrt{3} \\ -10 - j10\sqrt{3} \\ -5 - j5\sqrt{3} \end{bmatrix}, \text{ which gives } d = \begin{bmatrix} -5 + j5\sqrt{3} \\ -5 + j5\sqrt{3} \\ 10 \\ 10 \\ -5 - j5\sqrt{3} \\ -5 - j5\sqrt{3} \end{bmatrix}.
$$

For the line:

$$
c = \begin{bmatrix} 0 \\ -10 \\ -20 \\ -30 \\ -40 \\ -50 \end{bmatrix}, \text{ which gives } d = \begin{bmatrix} 10 \\ 10 \\ 10 \\ 10 \\ 10 \\ -50 \end{bmatrix}.
$$

Simulations are shown in Fig. 4.9. △

We conclude this section with a remark about invariance of formations. The controller (4.12) relies on a notion of formation that is invariant under translation. Indeed, two formation specifications c and $c + h\mathbf{1}$ give the same controller because $d = Lc = L(c + h\mathbf{1})$. However, rotating the formation does not give the same controller, so our notion of formation is not invariant under rotations. This is a little odd, as it would make more sense to stabilize formations modulo translations *and* rotations. To do that, a different representation of formation specifications is required, for instance in terms of desired distances between robots. When characterizing formations in terms of relative distances, the notion of formation rigidity comes into play. This topic is addressed in [20].

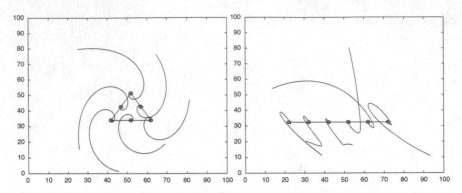

Fig. 4.9 Triangle and line formations for six integrator points in cyclic pursuit

4.6 Infinitely Many Robots

In Sect. 4.2 we considered cyclic pursuit of a finite number of integrator points. We found that the centroid remains stationary and the robots asymptotically rendezvous at the centroid. In this section, primarily out of mere curiosity, we see if and how that might be extended to infinitely many points.

There have been several papers written that consider infinitely many vehicles. For example, in 1999 Swaroop and Hedrick [43] introduced and studied the concept of string stability. Consider a one-dimensional straight road, with all traffic moving left to right as depicted in Fig. 4.10. The car shown farthest to the right is the leader— there is none to the right of it but there are infinitely many to its left. Suppose the drivers are controlling their speeds according to some rule. For example, each car accelerates or decelerates by a certain amount that depends on the gap between it and the car ahead. Suppose the leader brakes suddenly. Is there any collision in the convoy? If not, the convoy is said to possess *string stability*. Swaroop and Hedrick studied this concept.

The practical value of an infinite string may be questionable, although the approach may be useful when the number of robots is very large. Justh and Krishnaprasad [18] posed a problem having a continuum of masses, which is even more abstract.

Let $\mathbb{Z}$ denote the set of integers—negative, zero, and positive. Consider infinitely many integrator points:

$$\dot{z}_i = u_i, \quad i \in \mathbb{Z}.$$

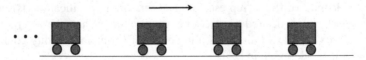

Fig. 4.10 Infinitely many cars moving left to right in a convoy

We cannot prescribe cyclic pursuit because there are not first and last robots. However, we can propose serial pursuit:

$$u_i = z_{i+1} - z_i.$$

Then

$$\dot{z}_i = z_{i+1} - z_i, \quad i \in \mathbb{Z}. \tag{4.13}$$

Bring in the infinite vector z and the infinite matrix U:

$$
z = \begin{bmatrix} \vdots \\ z_{-2} \\ z_{-1} \\ z_0 \\ z_1 \\ z_2 \\ \vdots \end{bmatrix}, \quad
U = \begin{bmatrix}
\vdots & \vdots & \vdots & \vdots & \vdots \\
\cdots & 0 & 1 & 0 & 0 & 0 & \cdots \\
\cdots & 0 & 0 & 1 & 0 & 0 & \cdots \\
\cdots & 0 & 0 & 0 & 1 & 0 & \cdots \\
\cdots & 0 & 0 & 0 & 0 & 1 & \cdots \\
\cdots & 0 & 0 & 0 & 0 & 0 & \cdots \\
\vdots & \vdots & \vdots & \vdots & \vdots
\end{bmatrix}.
$$

Then Eq. (4.13) can be written simply as

$$\dot{z} = Mz,$$

where $M = U - I$. The problem is to find if and when all the components of $z(t)$ converge to the same point. The answer turns out to be *yes*, and the rendezvous point is the origin, if the initial locations are square-summable:

$$\sum_{i \in \mathbb{Z}} |z_i(0)|^2 < \infty.$$

It is interesting that the centroid of such a set of initial locations is 0, that is,

$$\lim_{N \to \infty} \frac{1}{2N+1} \sum_{i=-N}^{N} z_i(0) = 0.$$

This is left as a calculus exercise.

Theorem 4.5 *If the components of $z(0)$ are square-summable, then all the components of $z(t)$ converge to 0 as t tends to infinity.*

A proof of this theorem is omitted, as it would take us too far afield. The interested reader is referred to Theorems 2 and 4 of [13].

The theorem is interesting for at least two reasons. The first is that the proof for the finite number of robots case, i.e., cyclic pursuit, does not carry over to the infinite robots case. To appreciate that, recall that the spectrum of $U - I$, that is, the set of eigenvalues, lies on the circle with centre -1 and radius 1; see Fig. 4.11. As we saw

Fig. 4.11 The eigenvalues
of $U - I$ for 8 robots in
cyclic pursuit

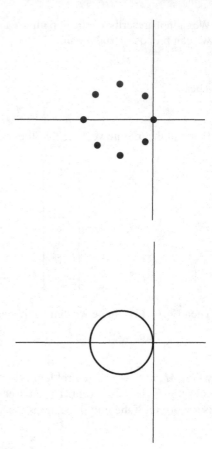

Fig. 4.12 The eigenvalues
of $U - I$ for infinitely many
robots

in the proof of Theorem 4.1, rendezvous in this case follows since, in the eigenvector expansion of $z(t)$, the seven terms for the left half-plane eigenvalues go to zero. The term for the zero eigenvalue is stationary. An eigenvector for this zero eigenvalue is **1**, the vector of all 1's. By contrast, for infinitely many robots the spectrum of $U - I$ is as is shown in Fig. 4.12. The spectrum is a circle with centre -1 and radius 1. Consequently, one cannot argue just from the spectrum that $z(t)$ decomposes into the direct sum of two parts, one convergent to zero and the other stationary.

The second reason why the theorem is interesting is that the result is not true if the condition "the components of $z(0)$ are square-summable" is replaced by "the components of $z(0)$ are bounded." For the consequence of this, imagine cars moving one way along an infinite road, as in Fig. 4.13. Suppose the cars are numbered by the integers and car n is travelling at speed $v_n(t)$. Suppose each car speeds up or slows down in order to maintain a constant distance from the car ahead of it. And suppose we, as control theorists, wish to know if this traffic flow is stable. Stable means that all cars will return to their original velocities after a perturbation. A reasonable perturbation is a small jump in $v_n(0)$. Let $v(t)$ denote the infinite vector with

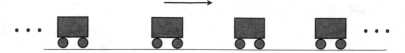

Fig. 4.13 Infinitely many cars

components $v_n(t)$. What is a reasonable notion for a small $v(0)$? From Theorem 4.5, we might expect that if $v(0)$ is square-summable, then the cars would return to their equilibrium speeds. But if all the speeds are perturbed by, say, 1 %, so that $v(0)$ is only bounded and not square summable, then we are not guaranteed a return to equilibrium.

4.7 On Digital Implementation of Controllers

As stated elsewhere, all systems in this book, including robots, have been modeled as operating in continuous time. Digital controllers, however, are typically used. In this brief section we discuss the ramifications of this setup.

Consider the simplest of all possible cyclic pursuits – two integrator points:

$$\dot{z}_1 = u_1 = z_2 - z_1, \quad \dot{z}_2 = u_2 = z_1 - z_2.$$

The associated block diagram is shown in Fig. 4.14. The equation $\dot{z}_1 = u_1$ says that z_1 is the integral of u_1; likewise for z_2 and u_2. Hence the integral signs inside the blocks. Consistent with this view, we regard an onboard controller to be a continuous-time system too when viewed from its input to its output. But it is implemented by two physical subsystems: a sampler S that converts a continuous-time signal into a discrete-time signal, followed by a discrete-to-continuous subsystem H, typically a zero-order hold. See Fig. 4.15. In this figure, since $u_1(t)$ and $u_2(t)$ are outputs of zero-order holds, they are constant between sampling times. Let us denote the sampling times as kT, k an integer.[3] The integral over a period of width T of a constant $u_1(kT)$ equals $Tu_1(kT)$. In this way we get that the sampled-data system in Fig. 4.15 has the discrete-time model

$$z_1[(k + 1)T] = z_1(kT) + T[z_2(kT) - z_1(kT)] \tag{4.14}$$

$$z_2[(k + 1)T] = z_2(kT) + T[z_1(kT) - z_2(kT)]. \tag{4.15}$$

It is important to realize in Fig. 4.15 that there are two physical digital controllers, and hence two physical samplers S. For Eqs. (4.14) and (4.15) to be valid, the two samplers S must be synchronized somehow, by a centralized clock and communication system.

---

[3]In this section alone, to avoid possible confusion the notation kT, k an integer, is used instead of tT, t an integer.

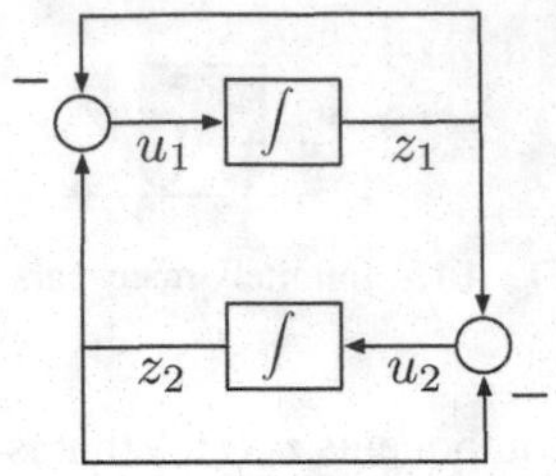

Fig. 4.14 Simple cyclic
pursuit

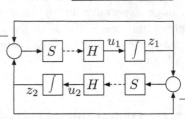

Fig. 4.15 After replacing
the unity controller by a
sample and hold. A *dotted
arrow* is a discrete-time
signal

So the system is not really distributed. Lack of synchroneity would lead to jitter, which
could alternatively be modelled. (It is interesting to note that system (4.14), (4.15) is
unstable for large enough T.)

As a concluding observation, we note that many distributed robotics articles, for
example [16] by Jadbabaie et al., begin with discrete-time models. Implicit in these
articles is the assumption that local clocks are synchronized by a global network.
A justification for this assumption is rarely given.

Chapter 5
The Rendezvous Problem: Limited Camera Range

5.1 Introduction

The theorems in Jadbabaie et al. [16] rely on the assumption that, for a fixed initial condition, the time dependent visibility graph satisfies a connectivity property. As we discussed in Chap. 3, for a sensible model of limited visibility, the graph is state dependent, not time dependent. To see this, let us return to our unicycle robots. Suppose robot i has an omnidirectional camera of range d_i. The set of visible neighbours of robot i is

$$\mathcal{N}_i(z) = \{j : j \in \{1, \ldots, n\}, \quad |z_j - z_i| \leq d_i\}.$$

Notice that, since $d_i \geq 0$, every robot is a neighbour of itself, and so $i \in \mathcal{N}_i(z)$. Then there is an edge in the visibility graph from i to j iff $j \in \mathcal{N}_i(z)$, and therefore the visibility graph is a function of z, $\mathcal{G}(z)$.

The stability analysis in the presence of state dependent graphs is quite difficult. Most researchers avoid this difficulty by assuming, as Jadbabaie et al. do, that the visibility graph is time dependent. There is, however, a clever control law that solves the rendezvous problem with limited camera range for kinematic points: the **circumcentre control law**. We turn to it next.

Example 5.1 Consider six robots, with omnidirectional cameras of identical ranges, positioned at $t = 0$ as shown in Fig. 5.1. The disks show the fields of view for robots 1 and 2. Since each camera has the same range, the visibility graph is undirected—see Fig. 5.2. Thus the neighbour sets at $t = 0$ are

$$\mathcal{N}_1 = \{1, 2\}, \quad \mathcal{N}_2 = \{1, 2, 4\}, \quad \mathcal{N}_3 = \{3, 4\},$$

$$\mathcal{N}_4 = \{2, 3, 4, 5\}, \quad \mathcal{N}_5 = \{4, 5, 6\}, \quad \mathcal{N}_6 = \{5, 6\}.$$

The circumcentre control law is defined as follows: Robot 1 has two neighbours, robot 1 itself and robot 2. Let $\mathcal{Z}_1$ denote the set of positions of the two neighbours,

© The Author(s) 2016
B.A. Francis and M. Maggiore, *Flocking and Rendezvous in Distributed Robotics*, SpringerBriefs in Control, Automation and Robotics, DOI 10.1007/978-3-319-24729-8_5

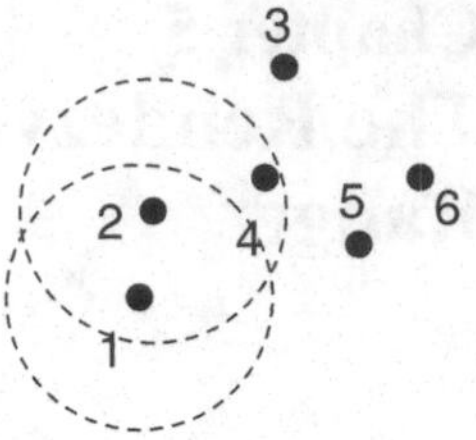

Fig. 5.1 Example of six
robots. The *disks* show the
regions of view of the
cameras on robots 1 and 2

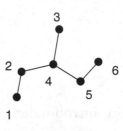

Fig. 5.2 The visibility graph

Fig. 5.3 How robot 1 moves

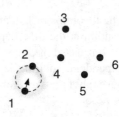

Fig. 5.4 How robot 2 moves

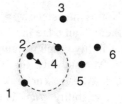

$\{z_1, z_2\}$, and let c_1 denote the *circumcentre* of $\mathcal{Z}_1$—the centre of the smallest circle containing $\mathcal{Z}_1$. Then set $u_1 = c_1 - z_1$—see Fig. 5.3. (In the picture, the little arrow is u_1 translated from the origin to z_1.) So robot 1 moves towards the centre at $t = 0$: $\dot{z}_1 = c_1 - z_1$. Actually, in this case where 1 sees only 2, clearly $c_1 = (z_1 + z_2)/2$, so at $t = 0$

$$\dot{z}_1 = \frac{1}{2}(z_2 - z_1).$$

Similarly, let c_2 denote the circumcentre of the set $\{z_j : j \in \mathcal{N}_2\}$ and define $u_2 = c_2 - z_2$—see Fig. 5.4. And so on.

These control laws can actually be implemented using onboard cameras, that is, relative positions, by translation. For example, for robot 2, the relative positions $\{z_1 - z_2, z_4 - z_2\}$ are sensed. Let $\mathcal{Z}'_2$ denote the set of points $\{0, z_1 - z_2, z_4 - z_2\}$

(the translate of $\{z_2, z_1, z_4\}$ by $-z_2$), and let c_2' denote the circumcentre of $\mathcal{Z}_2'$. Then define $u_2 = c_2'$.

Let us look at u_1 again. The set $\mathcal{Z}_1$ equals $\{z_1, z_2\}$ and so the circumcentre c_1 of $\mathcal{Z}_1$ is a function of z; write $c_1(z)$. It turns out that $c_1(z)$ is continuous in z, but not Lipschitz continuous—see below. In this way, the robots' motions are governed by the coupled equations

$$\dot{z}_1 = u_1(z) = c_1(z) - z_1$$

$$\vdots$$

$$\dot{z}_6 = u_6(z) = c_6(z) - z_6,$$

or in aggregate form $\dot{z} = u(z)$, where the vector field $u(z)$ is only continuous, not Lipschitz. Thus uniqueness of a solution is not guaranteed. In what follows, a statement about a solution should be interpreted as applying to all solutions if indeed there is more than one. $\triangle$

5.2 General Results

The fact that the circumcentre control law is not a Lipschitz continuous function causes difficulty in its use, as we will see later. So it is perhaps of interest to see a proof [3].

Lemma 5.1 *The circumcentre control law is not Lipschitz continuous.*

Proof Construct three points $\{p_1, p_2, p_3\}$ and their circumcentre c, and three perturbed points $\{p_1, p_2', p_3'\}$ and their circumcentre c'—see Fig. 5.5. Define the vectors

$$p = (p_1, p_2, p_3), \quad p' = (p_1, p_2', p_3').$$

Fig. 5.5 Proving non-Lipschitz

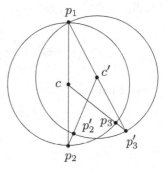

Fig. 5.6 Proving
non-Lipschitz (continued)

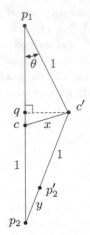

We will show that the ratio

$$\frac{|c - c'|}{\|p - p'\|}$$

is not bounded by a constant. This proves the mapping $p \mapsto c$ is not Lipschitz.

Let the radii of the circles be 1 and define $x = |c - c'|$, $y = |p_2 - p'_2|$. Since p_1 did not move and $|p_2 - p'_2| = |p_3 - p'_3|$,

$$\|p - p'\| = \sqrt{2}y.$$

Now look at Fig. 5.6. Define the angle θ. Then we have the lengths $\overline{p_1q} = \cos(\theta)$, $\overline{qc'} = \sin(\theta)$. Thus $\overline{qc} = 1 - \cos(\theta)$, so by Pythagoras on the small triangle qcc'

$$x^2 = (1 - \cos(\theta))^2 + \sin^2\theta = 2(1 - \cos(\theta)),$$

and therefore $\overline{qc} = x^2/2$.

By Pythagoras again on the triangle qcc', the length of $\overline{qc'}$ equals $x\sqrt{1 - \dfrac{x^2}{4}}$.
Finally, apply Pythagoras to triangle qp_2c':

$$(y + 1)^2 = \left(1 + \frac{x^2}{2}\right)^2 + x^2\left(1 - \frac{x^2}{4}\right) = 2x^2 + 1.$$

Thus we have

$$x = \sqrt{\frac{1}{2}y^2 + y},$$

and so

$$\frac{|c - c'|}{\|p - p'\|} = \frac{x}{\sqrt{2}y} = \frac{1}{2}\sqrt{1 + \frac{2}{y}}.$$

$\square$

The great thing about the circumcentre law is that it preserves connectivity of the visibility graph, unlike, say, heading for the centroid of the neighbour set. Thus we will have to assume only that the visibility graph is connected at $t = 0$. In fact, under the circumcentre control law, no links are dropped (though the distances between some neighbours may increase), so if $\mathcal{G}(z(0))$ is connected, then $\mathcal{G}(z(t))$ is connected for all $t > 0$. Of course, new links may form: As the robots rendezvous, eventually the graph becomes complete.

Lemma 5.2 *Under the circumcentre control law, over time no links are dropped in the visibility graph.*

Proof For this proof it is more convenient to view a robot position z_i as a vector in $\mathbb{R}^2$ instead of a complex number. Let $t \geq 0$ be arbitrary. Let $V_{ij}(z(t))$ denote the distance-squared between two neighbour robots i and j, and let $V(z(t))$ denote the maximum distance-squared between any two neighbours:

$$V(z(t)) = \max_i \max_{j \in \mathcal{N}_i(z(t))} V_{ij}(z(t)).$$

Let $\mathcal{I}(z(t))$ denote the set of pairs of indices where the maximum is attained; that is, $(i, j) \in \mathcal{I}(z(t))$ iff robots i and j are neighbours of maximum distance apart among all neighbours. Thus

$$V(z(t)) = \max_{(i,j) \in \mathcal{I}(z(t))} V_{ij}(z(t)).$$

We would like to show that $d/dt\, V(z(t)) \leq 0$. Unfortunately, $V(z(t))$ is not differentiable. We need some non-smooth analysis—the *upper Dini derivative*:

$$D^+ V(z(t)) = \lim \sup_{\tau \to 0^+} \frac{V(z(t + \tau)) - V(z(t))}{\tau}.$$

Then, it is a fact that

$$D^+ V(z(t)) = \max_{(i,j) \in \mathcal{I}(z(t))} \frac{d}{dt} V_{ij}(z(t)). \tag{5.1}$$

In this way we get

$$D^+ V(z(t)) = \max_{(i,j)\in\mathcal{I}(z(t))} \frac{d}{dt} \|z_i(t) - z_j(t)\|^2$$

$$= \max_{(i,j)\in\mathcal{I}(z(t))} 2\langle z_i(t) - z_j(t), \dot{z}_i(t) - \dot{z}_j(t)\rangle$$

$$= \max_{(i,j)\in\mathcal{I}(z(t))} 2\langle z_i(t) - z_j(t), u_i(t) - u_j(t)\rangle$$

$$= \max_{(i,j)\in\mathcal{I}(z(t))} 2\{\langle z_i(t) - z_j(t), u_i(t)\rangle + 2\langle z_j(t) - z_i(t), u_j(t)\rangle\}$$

$$\leq \max_{(i,j)\in\mathcal{I}(z(t))} 2\langle z_i(t) - z_j(t), u_i(t)\rangle$$

$$+ \max_{(i,j)\in\mathcal{I}(z(t))} 2\langle z_j(t) - z_i(t), u_j(t)\rangle.$$

To conclude that

$$D^+ V(z(t)) \leq 0, \tag{5.2}$$

we will show that

$$\max_{(i,j)\in\mathcal{I}(z(t))} \langle z_i(t) - z_j(t), u_i(t)\rangle \leq 0 \tag{5.3}$$

and

$$\max_{(i,j)\in\mathcal{I}(z(t))} \langle z_j(t) - z_i(t), u_j(t)\rangle \leq 0. \tag{5.4}$$

To illustrate the argument, suppose $(1, j) \in \mathcal{I}(z(t))$ for some j, that is, the maximum separation between robot neighbours occurs for robot 1 (and perhaps others). Suppose the neighbours of robot 1 are robots 2, 3, 4, 5 as in Fig. 5.7. See the circumcentre control vector u_1 (translated to z_1). The figure shows three neighbour robots—2, 3, and 4—on the smallest encompassing circle. Now in Fig. 5.8 construct the line as shown through z_1 perpendicular to u_1, and using this line as diameter, draw a second circle. In the shaded crescent there must be a neighbour of robot 1, for otherwise the encompassing circle in the figure would be smaller (in fact it would be the unshaded circle). Consider the robot in the shaded crescent that is maximum distance from robot 1; in the figure it is robot 3. The angle between the vectors u_1 and $z_1 - z_3$ is greater than $\pi/2$. Therefore

Fig. 5.7 Proving no links are dropped

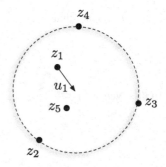

Fig. 5.8 Proving no links are dropped (continued)

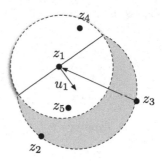

$$\langle z_1(t) - z_3(t), u_1(t) \rangle \le 0$$

and so

$$\max_{(1,j) \in \mathcal{I}(z(t))} \langle z_1(t) - z_j(t), u_1(t) \rangle \le 0.$$

This proves (5.3), and (5.4) follows from this.

Finally, from (5.1) and (5.2), if two neighbours i and j are maximum distance apart (among all neighbours), then $d/dt\, V_{ij}(z(t)) \le 0$ and so the distance between them is non-increasing. □

Here is the main result that the circumcentre control law solves the rendezvous problem:

Theorem 5.1 *Suppose $z(0)$ is such that $\mathcal{G}(z(0))$ is connected. Under the circumcentre control law, the robots rendezvous.*

The proof uses LaSalle's theorem. Here we want to discuss the ideas without the details.

Ideas for a Proof We are given that $\mathcal{G}(z(0))$ is connected. By Lemma 5.2, $\mathcal{G}(z(t))$ is connected for all $t > 0$. Now $\mathcal{G}(z(t))$ is either fixed or it is not. Suppose not. Then at some time a new link appears (no link is dropped). After this, $\mathcal{G}(z(t))$ is either fixed or it is not. Suppose not. Then at some time, another new link appears. Since there are only finitely many nodes, this process must stop. Thus we may assume without loss of generality that $\mathcal{G}(z(t))$ is fixed and connected for all $t \ge 0$. (We do not assume the graph is complete, but it must actually be so, since the robots rendezvous.)

Bring in the example in Fig. 5.9 for illustrative purposes. The picture in Fig. 5.10 shows the constellation at $t = 0$, its convex hull $\mathrm{co}\{z(0)\}$, and the instantaneous velocities $u_i(z(0))$ of the robots at the vertices. Even though the vector fields $u_i(z(0))$ point into $\mathrm{co}\{z(0)\}$, we cannot conclude that $z_i(t) \in \mathrm{co}\{z(0)\}$ because we do not have Lipschitz continuity. So it is problematic to prove even that a solution $z(t)$ is bounded.

Let a be an arbitrary point in the plane and define the function $V^a(z)$ to be the distance squared from a to the farthest z_i as in Fig. 5.11. (Again, we take the plane $\mathbb{R}^2$.) Assume for simplicity that the farthest-away robot does not change, that it is always robot 3. Then $V^a(z)$ is differentiable and

Fig. 5.9 Illustration

Fig. 5.10 Illustration
(continued)

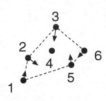

Fig. 5.11 Illustration
(continued)

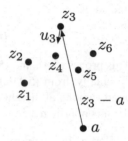

$$\frac{d}{dt} V^a(z(t)) = \frac{d}{dt} \|z_3(t) - a\|^2$$
$$= 2\langle u_3(z(t)), z_3(t) - a\rangle.$$

From the vector orientations in the figure, $\langle u_3(z(t)), z_3(t) - a\rangle \le 0$. Thus $V^a(z(t))$ is nonincreasing, and this kind of argument shows that $z(t)$ is defined for all $t > 0$ and is bounded.

Now invoke LaSalle's theorem. The solution converges to the largest invariant manifold $\mathcal{M}$ in $\{z : \dot{V}^a(z) = 0\}$. To see what this manifold is, let $z(0) \in \mathcal{M}$. Continuing with the assumption that the farthest-away $z_i(0)$ from a is $z_3(0)$, we have that $u_3(z(0))$ and $z_3(0) - a$ are orthogonal. Looking at the figure we conclude that $u_3(z(0)) = 0$; for if $u_3(z(0)) \ne 0$ then $z_4(0)$, the only neighbour of $z_3(0)$, is farther from a than is $z_3(0)$. Since $u_3(z(0)) = 0$, then z_3 and z_4 must be collocated at $t = 0$. If $z_2(0)$ and $z_5(0)$, the neighbours of $z_4(0)$, are not also collocated with $z_4(0)$, then $z_4(t)$ will move away from $z_3(0)$, which is impossible since $\mathcal{M}$ is invariant. Using this kind of argument, one can prove that for $z \in \mathcal{M}$, all z_i are equal.

A rigorous proof is considerably more complicated since $V^a(z)$ is not actually differentiable.

5.3 Numerical Issues

We have seen that the rendezvous problem with range-limited cameras, idealized to have perfect vision up to one range and zero vision beyond this range, has the interesting solution of a circumcentre control law. In this section we discuss what would be involved if we actually wanted to implement this control law.

Consider a set of distinct points, $p_1, \ldots, p_n$, in $\mathbb{C}$ and consider their circumcircle, $\mathcal{C}$. Either there are two points on $\mathcal{C}$ diametrically opposite, in which case the centre is easily calculated as the midpoint between them, or there are three points on $\mathcal{C}$ spanning an arc of more than π radians; in this case the centre lies within the triangle formed from the three points. How to find these boundary points is not discussed here.

We turn to the problem of computing the centre, c, of the circle in the latter case. The setup is shown in Fig. 5.12, where the three points lie on a circle and the centre of the circle lies inside the triangle. The centre is in the convex hull of p_0, p_1, p_2 and therefore can be parametrized as

$$c = \lambda_0 p_0 + \lambda_1 p_1 + \lambda_2 p_2, \tag{5.5}$$

where $\lambda_i \geq 0$ and

$$\lambda_0 + \lambda_1 + \lambda_2 = 1. \tag{5.6}$$

To see this, shift p_0 to the origin. Then $c - p_0$ is in the convex hull (line) of two points, $\alpha_1(p_1 - p_0)$, $\alpha_2(p_2 - p_0)$; see Fig. 5.13, Thus

$$c - p_0 = \lambda\alpha_1(p_1 - p_0) + (1 - \lambda)\alpha_2(p_2 - p_0),$$

and hence

$$c = [1 - \lambda\alpha_1 - (1 - \lambda)\alpha_2]p_0 + \lambda\alpha_1 p_1 + (1 - \lambda)\alpha_2 p_2,$$

which has the form (5.5).

Fig. 5.12 Computing the circumcentre

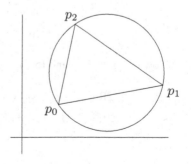

Fig. 5.13 Computing the
circumcentre (continued)

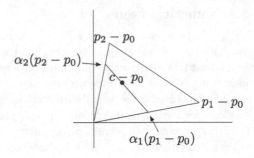

The centre satisfies the two equations

$$|p_2 - c| = |p_0 - c|$$
$$|p_1 - c| = |p_0 - c|.$$

Thus

$$(p_2 - c)(\bar{p}_2 - \bar{c}) = (p_0 - c)(\bar{p}_0 - \bar{c})$$
$$(p_1 - c)(\bar{p}_1 - \bar{c}) = (p_0 - c)(\bar{p}_0 - \bar{c})$$

and hence

$$|p_2|^2 - 2\mathrm{Re}(c\bar{p}_2) + |c|^2 = |p_0|^2 - 2\mathrm{Re}(c\bar{p}_0) + |c|^2$$
$$|p_1|^2 - 2\mathrm{Re}(c\bar{p}_1) + |c|^2 = |p_0|^2 - 2\mathrm{Re}(c\bar{p}_0) + |c|^2.$$

This gives

$$|p_2|^2 - 2\mathrm{Re}(c\bar{p}_2) = |p_0|^2 - 2\mathrm{Re}(c\bar{p}_0)$$
$$|p_1|^2 - 2\mathrm{Re}(c\bar{p}_1) = |p_0|^2 - 2\mathrm{Re}(c\bar{p}_0),$$

or

$$2\mathrm{Re}[c(\bar{p}_0 - \bar{p}_2)] = |p_0|^2 - |p_2|^2$$
$$2\mathrm{Re}[c(\bar{p}_0 - \bar{p}_1)] = |p_0|^2 - |p_1|^2,$$

Now bring in (5.5) and (5.6):

$$2\mathrm{Re}[(\lambda_0 p_0 + \lambda_1 p_1 + \lambda_2 p_2)(\bar{p}_0 - \bar{p}_2)] = |p_0|^2 - |p_2|^2$$
$$2\mathrm{Re}[(\lambda_0 p_0 + \lambda_1 p_1 + \lambda_2 p_2)(\bar{p}_0 - \bar{p}_1)] = |p_0|^2 - |p_1|^2$$
$$\lambda_0 + \lambda_1 + \lambda_2 = 1.$$

Thus

$$\lambda_0 2\text{Re}[p_0(\bar{p}_0 - \bar{p}_2)] + \lambda_1 2\text{Re}[p_1(\bar{p}_0 - \bar{p}_2)] + \lambda_2 2\text{Re}[p_2(\bar{p}_0 - \bar{p}_2)] = |p_0|^2 - |p_2|^2$$
$$\lambda_0 2\text{Re}[p_0(\bar{p}_0 - \bar{p}_1)] + \lambda_1 2\text{Re}[p_1(\bar{p}_0 - \bar{p}_1)] + \lambda_2 2\text{Re}[p_2(\bar{p}_0 - \bar{p}_1)] = |p_0|^2 - |p_1|^2$$
$$\lambda_0 + \lambda_1 + \lambda_2 = 1.$$

This can be written as

$$\begin{bmatrix} 2\text{Re}[p_0(\bar{p}_0 - \bar{p}_2)] & 2\text{Re}[p_1(\bar{p}_0 - \bar{p}_2)] & 2\text{Re}[p_2(\bar{p}_0 - \bar{p}_2)] \\ 2\text{Re}[p_0(\bar{p}_0 - \bar{p}_1)] & 2\text{Re}[p_1(\bar{p}_0 - \bar{p}_1)] & 2\text{Re}[p_2(\bar{p}_0 - \bar{p}_1)] \\ 1 & 1 & 1 \end{bmatrix} \begin{bmatrix} \lambda_0 \\ \lambda_1 \\ \lambda_2 \end{bmatrix}$$
$$= \begin{bmatrix} |p_0|^2 - |p_2|^2 \\ |p_0|^2 - |p_1|^2 \\ 1 \end{bmatrix}.$$

The solution of this equation gives c via (5.5).

What about numerical sensitivity of this procedure? Consider the example in Fig. 5.14. The points form an isosceles triangle: $p_0 = 0$, $p_1 = je^{-j\varepsilon}$, $p_2 = je^{j\varepsilon}$. The equation for the λ_i's is

$$\begin{bmatrix} 0 & -2\text{Re}[p_1\bar{p}_2] & -2\text{Re}[p_2\bar{p}_2] \\ 0 & -2\text{Re}[p_1\bar{p}_1] & -2\text{Re}[p_2\bar{p}_1] \\ 1 & 1 & 1 \end{bmatrix} \begin{bmatrix} \lambda_0 \\ \lambda_1 \\ \lambda_2 \end{bmatrix} = \begin{bmatrix} -|p_2|^2 \\ -|p_1|^2 \\ 1 \end{bmatrix},$$

and thus

$$\begin{bmatrix} 0 & 2\text{Re}[je^{-j\varepsilon}je^{-j\varepsilon}] & -2 \\ 0 & -2 & 2\text{Re}[je^{j\varepsilon}je^{j\varepsilon}] \\ 1 & 1 & 1 \end{bmatrix} \begin{bmatrix} \lambda_0 \\ \lambda_1 \\ \lambda_2 \end{bmatrix} = \begin{bmatrix} -1 \\ -1 \\ 1 \end{bmatrix},$$

or

$$\begin{bmatrix} 0 & -2\text{Re}[e^{-j2\varepsilon}] & -2 \\ 0 & -2 & -2\text{Re}[e^{j2\varepsilon}] \\ 1 & 1 & 1 \end{bmatrix} \begin{bmatrix} \lambda_0 \\ \lambda_1 \\ \lambda_2 \end{bmatrix} = \begin{bmatrix} -1 \\ -1 \\ 1 \end{bmatrix},$$

Fig. 5.14 Sensitivity of computing the circumcentre

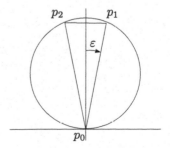

or finally

$$\begin{bmatrix} 0 & 2\cos(2\varepsilon) & 2 \\ 0 & 2 & 2\cos(2\varepsilon) \\ 1 & 1 & 1 \end{bmatrix} \begin{bmatrix} \lambda_0 \\ \lambda_1 \\ \lambda_2 \end{bmatrix} = \begin{bmatrix} 1 \\ 1 \\ 1 \end{bmatrix}.$$

As $\varepsilon \to 0$, the left-hand matrix becomes singular, though $\lambda_0, \lambda_1, \lambda_2$ converge respectively to $1/2, 1/4, 1/4$. Thus the linear equation to be solved can become ill-conditioned. A sensible thing to do if you have two points very close together (like p_1, p_2) is to ignore one of them.

Finally, consider $n \geq 2$ point robots moving according to the circumcentre control law. Assuming the initial visibility graph has a globally reachable node, the robots rendezvous asymptotically. Thus eventually, say at time t_1, they are all visible to each other and there is only one circumcircle. On the time interval $[t_1, \infty)$, the circumcentre is stationary.

To see this, suppose three robots, z_1, z_2, z_3, are on the circumcircle, all the others being inside. We have for each robot that $\dot{z}_i = c - z_i$. Thus

$$|\dot{z}_1| = |\dot{z}_2| = |\dot{z}_3| =: s_{max},$$

the maximum speed, while for every other i, $|\dot{z}_i| < s_{max}$. It follows that z_1, z_2, z_3 move radially toward c at the same speed and that they remain on the circumcircle as it shrinks. The centre of the circumcircle remains stationary.

Chapter 6
Introduction to Flying Robots

6.1 Introduction

In this chapter we discuss helicopter flying robots. Unlike airplanes, helicopters have the ability to hover at a fixed position in space. Their source of propulsion is a collection of one or more rotors that can be thought of as rotating wings—see Fig. 6.1. Similar to a wing, the relative velocity between the rotor blades and the surrounding air generates a force transversal to the rotor plane. This force can be reasonably assumed to be orthogonal to the rotor plane and to have a magnitude proportional to the squared speed of rotation.

A rotor alone is not sufficient to produce stable flight. Some mechanism is required to steer the helicopter, for otherwise one would have no way to affect the direction of flight. Additionally, by Newton's third law, a torque on the rotor shaft will cause an opposite torque on the helicopter body, causing the helicopter to spin about the rotor axis. One must therefore devise a way to counteract this spin and to produce steering torques. There are many ways to achieve these desired properties; we will mention a few.

6.1.1 Common Flying Robots

A ducted fan aircraft, schematically depicted in Fig. 6.2, has only one rotor and two pairs of ailerons underneath it. Each aileron pair works like the ailerons on the wings of an airplane, producing torques about two orthogonal axes. Together, the four ailerons produce torques around three independent axes, and can be used to steer the aircraft while preventing it from spinning.

A conventional helicopter has two rotors. The main rotor generates a lift force while the tail rotor stabilizes the spin. The main rotor is mounted on a swash plate that varies the pitch of the rotor blades to produce appropriate roll and pitch torques.

© The Author(s) 2016
B.A. Francis and M. Maggiore, *Flocking and Rendezvous in Distributed Robotics*, SpringerBriefs in Control, Automation and Robotics, DOI 10.1007/978-3-319-24729-8_6

Fig. 6.1 The rotor of a
helicopter, rotating with
speed ω and producing a lift
force with magnitude $k\omega^2$

Fig. 6.2 Ducted fan aircraft

Fig. 6.3 Coaxial helicopter

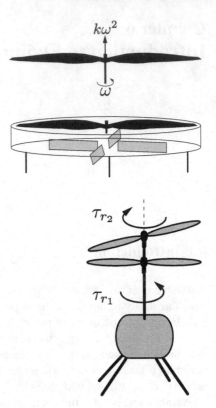

The configuration of a conventional helicopter is not ideal for a miniature flying robot, since the energy expended by the tail rotor is not used to generate lift. A more energy efficient configuration is that of a co-axial helicopter, displayed in Fig. 6.3. This helicopter has two rotors mounted on a common axis and rotating in opposite directions. In this configuration, both rotors contribute to producing a lift force. Moreover, if τ_{r_1} and τ_{r_2} denote the torques applied by motors to the two rotors—see Fig. 6.3—then the helicopter body is subjected to a differential torque, $\tau_{r_1} - \tau_{r_2}$, about the rotor axis, which can be used to prevent the helicopter from spinning. Like conventional helicopters, coaxial helicopters often use swash plates for steering.

Quadrotor helicopters (also called quadrocopters) are the most common flying robots. The structure of a quadrotor helicopter is shown in Fig. 6.4. It has four coplanar rotors. Viewed from the top, the rotor shafts are placed on the vertices of a square. Two rotors on a diagonal of the square rotate clockwise; the other two rotate counterclockwise. In Fig. 6.4, τ_{r_i} denotes the torque that the ith motor applies to rotor i, and f_{r_i} denotes the magnitude of the force generated by rotor i (τ_{r_i} and f_{r_i} are naturally related to one another; we will discuss this relationship later). As a result of this arrangement, three torques are applied about the body axes $\{b_1, b_2, b_3\}$ of Fig. 6.4. By Newton's third law, the torque $\tau_{r_1} + \tau_{r_2} - \tau_{r_3} - \tau_{r_4}$ is applied about axis b_3; forces f_{r_1} and f_{r_2} produce a moment about axis b_1, resulting in the torque

Fig. 6.4 Quadrotor
helicopter

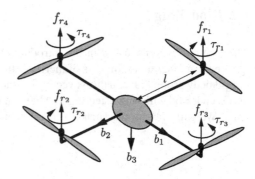

$l(f_{r_1} - f_{r_2})$; similarly, forces f_{r_3} and f_{r_4} produce the torque $l(f_{r_3} - f_{r_4})$ about axis b_2. Finally, the sum of the four forces $f_{r_1} + \cdots + f_{r_4}$ is the total vertical thrust on the helicopter.

6.1.2 Onboard Sensors

Every flying robot is equipped with an inertial measurement unit (IMU). IMUs typically contain three accelerometers, three rate-gyroscopes, and three magnetometers. The accelerometers measure the non-gravitational acceleration vector in a local coordinate frame (often the frame is printed on the chip), that is, the difference between the acceleration of the vehicle and the acceleration due to gravity. The rate-gyroscopes measure the roll, pitch, and yaw angle rates of the robot. Finally, the magnetometers measure the coordinates of the earth's magnetic field vector in the local frame.

In addition to IMUs, flying robots may have one or more ultrasonic sensors to detect proximity to objects or to the ground, a barometric pressure sensor, one or more cameras (one pointing along the helicopter's heading axis and sometimes one pointing downward), and a GPS sensor.

Now consider a collection of flying robots, each carrying a camera and a marker. If robot j's marker is in the field of view of robot i's camera, then from the dimension and position of the marker in the camera image, robot i can deduce its relative displacement to robot j in its own local frame. Moreover if the marker is suitably designed, from the camera image it is possible to deduce the relative rotation between the camera frame and the marker. In conclusion, we may assume that, in addition to the IMU data, vehicle i has access to the relative displacement and relative orientation of vehicle j with respect to vehicle i's local frame.

6.2 Modelling

The flying robots discussed in the introduction have something in common. If we ignore the masses and moments of inertia of the rotors, each robot may be viewed as a rigid body propelled by a thrust vector that has constant direction in the body frame and is endowed with some steering mechanism that induces torques about three body axes. We will now model such a general setup, beginning with the simpler planar case.

6.2.1 2D Flying Robot

We begin with the simplified setup of Fig. 6.5, in which the robot flies in a horizontal plane, propelled by a thrust vector f parallel to its heading. The state of the robot is the position $x = (x_1, x_2)$ of its centre of mass in an inertial coordinate frame, the velocity $\dot{x}$, the heading angle θ, and the angular speed $\dot{\theta}$. There are two control inputs: the magnitude, u, of the thrust vector and the torque, τ, about the axis coming out of the page. The thrust vector f in inertial coordinates is given by

$$f = u \begin{bmatrix} \cos(\theta) \\ \sin(\theta) \end{bmatrix}.$$

Let m denote the mass of the robot. Then Newton's equation gives

$$m\ddot{x} = f.$$

Fig. 6.5 2D flying robot

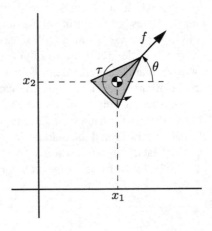

Fig. 6.6 Body frame

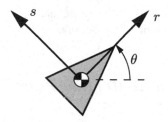

The torque τ causes the robot to rotate. Denoting by J the moment of inertia of the robot about the axis coming out of the page, computed with respect to the centre of mass, we have

$$J\ddot{\theta} = \tau.$$

The complete model is therefore

$$m\ddot{x} = u \begin{bmatrix} \cos(\theta) \\ \sin(\theta) \end{bmatrix}$$

$$J\ddot{\theta} = \tau.$$ (6.1)

There is a better way to represent model (6.1), one that lends itself to a 3D generalization. As we did in Chap. 2 for the unicycle model, we attach an orthonormal frame $\mathcal{B} = \{r, s\}$ to the robot as in Fig. 6.6, and we define the rotation matrix of frame $\mathcal{B}$ with respect to the inertial frame as

$$R := [r \ s] = \begin{bmatrix} \cos(\theta) & -\sin(\theta) \\ \sin(\theta) & \cos(\theta) \end{bmatrix}.$$

Then, as in Chap. 2, we have

$$\dot{R} = R \begin{bmatrix} 0 & -\omega \\ \omega & 0 \end{bmatrix} = RS(\omega).$$

The thrust vector f is parallel to the body frame axis r and has magnitude u,

$$f = uRe_1,$$

where $e_1 = [1 \ 0]^T$. In conclusion, we can rewrite model (6.1) as follows:

$$m\ddot{x} = uRe_1$$

$$\dot{R} = RS(\omega)$$ (6.2)

$$J\dot{\omega} = \tau.$$

We will now see that this model has a straightforward generalization to the three-dimensional setting.

6.2.2 3D Flying Robot

Consider now the setup of Fig. 6.7, in which the robot flies in the three-dimensional Euclidean space. As before, we fix an inertial frame $\mathcal{I}$ and a body frame $\mathcal{B} = \{b_1, b_2, b_3\}$, both orthonormal and right-handed. We denote by $x = (x_1, x_2, x_3)$ the coordinates of the robot's centre of mass in frame $\mathcal{I}$. The body is propelled by a thrust vector f that is now assumed to point opposite to the body axis b_3. There are four control inputs: the magnitude u of the thrust vector and three torques τ_1, τ_2, τ_3 about the three body axes. As in the planar case, we define the **rotation matrix** of frame $\mathcal{B}$ with respect to frame $\mathcal{I}$,

$$R = [b_1 \ \ b_2 \ \ b_3].$$

We pause for a moment to highlight a notable property of the rotation matrix R. Consider a vector with coordinates (v_1, v_2, v_3) in frame $\mathcal{B}$—see Fig. 6.8. Now translate the tail of this vector to the origin of frame $\mathcal{I}$ and let (w_1, w_2, w_3) be the coordinates in frame $\mathcal{I}$ of this translated vector. What is the relationship between the two coordinate representations (v_1, v_2, v_3) and (w_1, w_2, w_3)? Bring in the unit vectors b_i and note that

$$\begin{bmatrix} w_1 \\ w_2 \\ w_3 \end{bmatrix} = v_1 b_1 + v_2 b_2 + v_3 b_3 = R \begin{bmatrix} v_1 \\ v_2 \\ v_3 \end{bmatrix}.$$

In conclusion, the matrix R can be used to change the coordinate representation of vectors between frames $\mathcal{B}$ and $\mathcal{I}$.

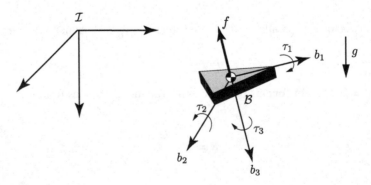

Fig. 6.7 3D flying robot

Fig. 6.8 Two coordinate representations in frames $\mathcal{I}$ and $\mathcal{B}$

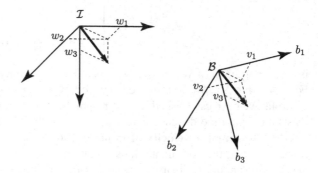

We now return to modelling. The thrust vector f is proportional to $-b_3$ and has magnitude u. Therefore in the coordinates of frame $\mathcal{I}$ we have

$$f = -u R e_3.$$

The gravity vector is parallel to the third axis of frame $\mathcal{I}$, and therefore its coordinate representation in frame $\mathcal{I}$ is $g e_3$. Newton's equation gives

$$m \ddot{x} = m g e_3 - u R e_3.$$

Next we need to model the rotation of the body. We begin with the observation that, since the columns of R form an orthonormal basis of $\mathbb{R}^3$, it holds that $R^T R = I_3$, where I_3 is the 3×3 identity matrix. Now suppose that the body undergoes a rotation, so that R varies with time, $R = R(t)$. It still holds that $R(t)^T R(t) = I_3$ for all t. Differentiating both sides of this identity with respect to t, we obtain

$$\dot{R}^T R + R^T \dot{R} = 0_{3 \times 3},$$

or

$$(R^T \dot{R})^T = -(R^T \dot{R}).$$

In other words, the matrix $R^T \dot{R}$ is skew-symmetric. Now, it is an easy exercise to show that an arbitrary 3×3 skew-symmetric matrix has the form

$$\begin{bmatrix} 0 & -\omega_3 & \omega_2 \\ \omega_3 & 0 & -\omega_1 \\ -\omega_2 & \omega_1 & 0 \end{bmatrix},$$

for suitable $\omega_1, \omega_2, \omega_3$. In fact, the set of 3×3 real skew-symmetric matrices

$$se(3) = \{S \in \mathbb{R}^{3 \times 3} : S^T = -S\}$$

is a subspace of $\mathbb{R}^{3 \times 3}$, and the map $\mathbb{R}^3 \to se(3)$,

$$\omega \mapsto S(\omega) := \begin{bmatrix} 0 & -\omega_3 & \omega_2 \\ \omega_3 & 0 & -\omega_1 \\ -\omega_2 & \omega_1 & 0 \end{bmatrix}$$

is an isomorphism of vector spaces. It is a fun exercise to prove the following fact: For any $\omega, v \in \mathbb{R}^3$, $S(\omega)v$ is the vector (or cross) product of ω and v, $S(\omega)v = \omega \times v$.

So far we have established that corresponding to a rotation of the frame $\mathcal{B}$, there exists a unique time-dependent vector $\omega(t)$ such that $R^T \dot{R} = S(\omega)$. The vector ω is called the **angular velocity** of frame $\mathcal{B}$ with respect to frame $\mathcal{I}$ represented in frame $\mathcal{B}$. Using the coordinate transformation property of R described earlier, one gets that the angular velocity in frame $\mathcal{I}$ is $R\omega$, but this vector will not be needed in the following development. Since $RR^T = I$, multiplying both sides of the identity $R^T \dot{R} = S(\omega)$ on the left by R we get

$$\dot{R} = RS(\omega).$$

This is an ordinary differential equation whose state is the rotation matrix R. This equation models the kinematics of rotation of the robot in terms of its body frame angular velocity.

Now we need to model the evolution of ω. Euler's second law states that the rate of change of the angular momentum in frame $\mathcal{I}$ is equal to the total external torque applied to the robot. Let J denote the inertia matrix of the robot in the coordinates of frame $\mathcal{B}$, defined with respect to its centre of mass. We assume that the robot is a rigid body, so J is a constant matrix. The angular momentum of the robot in frame $\mathcal{B}$ is the vector $J\omega$. The representation of this vector in frame $\mathcal{I}$ is $RJ\omega$. The robot is actuated by a torque vector $\tau = (\tau_1, \tau_2, \tau_3)$ in frame $\mathcal{B}$, or $R\tau$ in frame $\mathcal{I}$. Therefore, Euler's second law gives

$$\frac{d}{dt}(RJ\omega) = R\tau.$$

Multiplying both sides of the equation by R^T and using the identity $R^T R = I$, we get

$$R^T \frac{d}{dt}(RJ\omega) = \tau,$$

or

$$R^T(\dot{R}J\omega + RJ\dot{\omega}) = \tau.$$

Using the identity $\dot{R} = RS(\omega)$ and rearranging terms we obtain

$$J\dot{\omega} + S(\omega)J\omega = \tau.$$

Since $S(\omega)v = \omega \times v$, the term $S(\omega)J\omega$ equals $\omega \times (J\omega)$.

In conclusion, the model[1] of the 2D flying robot is

$$m\ddot{x} = mge_3 - uRe_3$$
$$\dot{R} = RS(\omega) \qquad (6.3)$$
$$J\dot{\omega} = -\omega \times (J\omega) + \tau.$$

The state is $(x, \dot{x}, R, \omega)$. There are four control inputs: the magnitude u of the thrust vector and the three torques $\tau = (\tau_1, \tau_2, \tau_3)$.

6.2.3 Special Case: Quadrotor Helicopters

A quadrotor helicopter can be regarded as a rigid body propelled by a force vector with constant direction in its body frame. Moreover as we have seen in Sect. 6.1, with a judicious choice of the rotor speeds one may induce desired torques about the three body axes. Quadrotor helicopters, therefore, fit within the class of robots modelled by (6.3). How are the controls (u, τ) in (6.3) related to the physical control inputs of a quadrotor helicopter? In this section we answer this question.

Consider the quadrotor helicopter of Fig. 6.9. The four rotors are driven by electric motors applying torques τ_{r_i}, $i = 1, \ldots, 4$. These can be regarded as the physical control inputs. We seek a relationship between these inputs and the controls (u, τ) of the model (6.3), also depicted in Fig. 6.9. To this end, denote by ω_{r_i} the angular speed of rotor i. Then it can be shown that each rotor is modelled by

$$J_{r_i}\dot{\omega}_{r_i} = -B\omega_{r_i}^2 + \tau_{r_i}, \quad i = 1, \ldots, 4,$$

where the term $-B\omega_{r_i}^2$ represents a torque due to the drag effects on the rotor blades. In practice the moment of inertia J_{r_i} is negligible, so it is common to use a singular perturbation argument and set $J_{r_i}\dot{\omega}_{r_i} = 0$, which gives $\tau_{r_i} = B\omega_{r_i}^2$. The lift force produced by rotor i is

$$f_{r_i} = k\omega_{r_i}^2 = \frac{k}{B}\tau_{r_i}. \qquad (6.4)$$

Recall from Sect. 6.1 that

$$u = f_{r_1} + f_{r_2} + f_{r_3} + f_{r_4}$$
$$\tau_1 = l(f_{r_1} - f_{r_2})$$
$$\tau_2 = l(f_{r_3} - f_{r_4})$$
$$\tau_3 = \tau_{r_1} + \tau_{r_2} - \tau_{r_3} - \tau_{r_4}.$$

---

[1]In the derivation of this model we have ignored drag and other aerodynamic effects. We have also ignored the dynamics inherent in the propulsion mechanism.

Fig. 6.9 The physical inputs τ_{r_i} of a quadrotor helicopter and the corresponding controls u, τ_1, τ_2, τ_3 used in model (6.3)

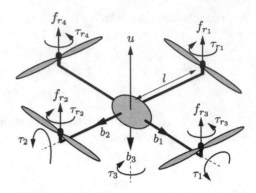

Using (6.4) we obtain an invertible feedback transformation $(\tau_{r_1}, \ldots, \tau_{r_4}) \mapsto (u, \tau)$:

$$
\begin{bmatrix} u \\ \tau_1 \\ \tau_2 \\ \tau_3 \end{bmatrix} = \frac{k}{B} \begin{bmatrix} 1 & 1 & 1 & 1 \\ l & -l & 0 & 0 \\ 0 & 0 & l & -l \\ \frac{B}{k} & \frac{B}{k} & -\frac{B}{k} & -\frac{B}{k} \end{bmatrix} \begin{bmatrix} \tau_{r_1} \\ \tau_{r_2} \\ \tau_{r_3} \\ \tau_{r_4} \end{bmatrix}. \tag{6.5}
$$

In conclusion, the model of a quadrotor helicopter with control inputs $(\tau_{r_1}, \ldots, \tau_{r_4})$ is given by (6.3) with (u, τ) given in (6.5).

6.3 Flocking of 2D Flying Robots

We now turn to the most basic coordination problem, flocking. We begin with the 2D case. Thus we have n robots modelled by

$$
\begin{aligned}
m_i \ddot{x}_i &= u_i R_i e_1 \\
\dot{R}_i &= R_i S(\omega_i) \quad i = 1, \ldots, n. \\
J_i \dot{\omega}_i &= \tau_i.
\end{aligned} \tag{6.6}
$$

A clarification about notation. From now on, a subscript i on a quantity indicates that the quantity pertains to the ith robot.[2] In particular, $\mathcal{B}_i$ is the body frame of robot i. We let $\chi_i := (x_i, \dot{x}_i, R_i, \omega_i)$ denote the state of the ith robot, and $\chi := (\chi_1, \ldots, \chi_n)$ denote the collective state.

As in Chap. 3, we indicate by $\mathcal{N}_i(\chi)$ the set of neighbours of robot i, and we rely on the visibility graph $\mathcal{G}(\chi)$ to keep track of who can see whom.

---

[2]This choice of notation creates a minor inconsistency with the previous section, where we have used, for instance, ω_1 to denote the first component of the angular velocity vector ω. From now on, ω_i will denote instead the angular velocity vector of robot i.

Fig. 6.10 Relative displacement and orientation between robots i and k as measured by robot i

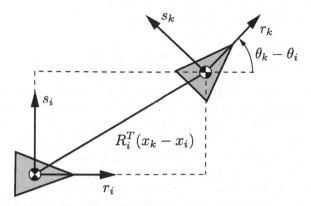

Before delving into control design, we need to clarify what is an admissible control for a flying robot. As discussed in Sect. 6.1, the onboard sensors of robot i can be used to deduce robot i's relative orientation with respect to any neighbour $k \in \mathcal{N}_i(\chi)$ and the relative displacement measured in the local frame $\mathcal{B}_i$. These quantities are illustrated in Fig. 6.10. We may also assume the time derivatives of the above quantities to be available for feedback. Moreover, one of the rate gyroscopes in the IMU measures ω_i. In conclusion, any control law (u_i, τ_i) will be called **admissible** for robot i if it is a locally Lipschitz function depending only on the quantities

$$\langle x_k - x_i, r_i \rangle, \quad \langle x_k - x_i, s_i \rangle, \quad \langle r_k, r_i \rangle, \quad \langle r_k, s_i \rangle,$$
$$\langle \dot{x}_k - \dot{x}_i, r_i \rangle, \quad \langle \dot{x}_k - \dot{x}_i, s_i \rangle, \quad \omega_i,$$

for each $k \in \mathcal{N}_i(\chi)$.

An equivalent representation of the measured quantities above is obtained by leveraging the coordinate transformation property of rotation matrices. For instance, the relative displacement between robots i and k measured in robot i's local frame is obtained by converting the inertial displacement $x_k - x_i$ to the coordinates of frame $\mathcal{B}_i$,

$$(R_i)^{-1}(x_k - x_i) = R_i^T(x_k - x_i).$$

The relative orientation between robot i and robot k is $R_i^T R_k$ (you can check that this matrix corresponds to a counterclockwise rotation by angle $\theta_k - \theta_i$). In essence, the measured quantities above can be represented as

$$R_i^T(x_k - x_i), \quad R_i^T(\dot{x}_k - \dot{x}_i), \quad R_i^T R_k, \quad \omega_i.$$

As in Chap. 3, the **flocking problem** for the robots in (6.6) is to find, if they exist, admissible controls (u_i, τ_i) so that there exists $\varepsilon > 0$ such that for all initial conditions satisfying

(i) $(\forall i, j = 1, \ldots, n) \, |\dot{x}_i(0) - \dot{x}_j(0)| < \varepsilon, \, |\theta_i(0) - \theta_j(0)| < \varepsilon, \, |\dot{\theta}_i(0)| < \varepsilon,$

(ii) $\mathcal{G}(\chi(0))$ is connected,

then

$$(\exists v_{ss}) \lim_{t \to \infty} \dot{x}_i(t) = v_{ss}, \quad \text{for all } i = 1, \dots, n.$$

In other words, the flying robots are required to move asymptotically along parallel straight lines whose slope depends on their initial conditions.

As for kinematic unicycles, there is currently no solution to the flocking problem. The main difficulty is the requirement of preserving the connectivity of the visibility graph, a problem of considerable difficulty even when the robots are modelled as kinematic integrators (see Chap. 5).

In the interest of a self-contained mathematical treatment, we make two assumptions. First, we assume that the visibility graph is constant (the set of neighbours of each robot does not change with time) and undirected (if robot i can see robot j, then robot j can see i). Second, we assume that each robot is affected by a drag force pointing opposite to its velocity vector:

$$m_i \ddot{x}_i = u_i R_i e_1 - b \dot{x}_i, \quad b > 0.$$

The above assumptions considerably simplify the flocking problem, for they allow us to discard the translational dynamics and focus on the subsystem

$$\dot{R}_i = R_i S(\omega_i)$$
$$J_i \dot{\omega}_i = \tau_i, \quad i = 1, \dots, n,$$

or, equivalently,

$$J_i \ddot{\theta}_i = \tau_i, \quad i = 1, \dots, n. \tag{6.7}$$

Indeed, set $u_1 = \cdots = u_n = \bar{u} > 0$ and suppose we were to design admissible controls τ_i making the angles θ_i converge to a common constant value θ_{ss}. Then the translational motion of each robot would be described by

$$m_i \frac{d}{dt} \dot{x}_i = -b \dot{x}_i + \bar{u} \begin{bmatrix} \cos(\theta_{ss}) \\ \sin(\theta_{ss}) \end{bmatrix} + \bar{u} \delta_i(t),$$

where $\delta_i(t) = (\cos(\theta_i(t)) - \cos(\theta_{ss}), \sin(\theta_i(t)) - \sin(\theta_{ss}))$ is a vanishing signal. The above differential equation is a stable linear time invariant system driven by an input signal converging to the constant vector. For this system we have $\dot{x}_i \to v_{ss}$, with $v_{ss} = \bar{u}/b(\cos(\theta_{ss}), \sin(\theta_{ss}))$, and the robots flock. The situation is illustrated in the block diagrams of Fig. 6.11.

To summarize, the flocking problem has been reduced to finding controls τ_i for system (6.7) making $(\theta_i, \dot{\theta}_i) \to (\theta_{ss}, 0)$, $i = 1, \dots, n$. An admissible control law for the reduced model (6.7) is one that relies only on $\theta_i - \theta_k$, $k \in \mathcal{N}_i$, and $\dot{\theta}_i$.

Our approach in the control design that follows will be to asymptotically stabilize the **flocking manifold**

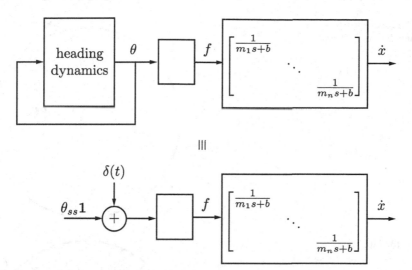

Fig. 6.11 Block diagram of 2D flying robots with synchronization of the heading angles. In the figure, f is the vector of thrust forces $f_i = u_i R_i e_1$

$$\mathcal{F} = \{(\theta_1, \ldots, \theta_n, \dot{\theta}_1, \ldots, \dot{\theta}_n) : \theta_1 = \cdots = \theta_n, \dot{\theta}_1 = \cdots = \dot{\theta}_n = 0\}, \qquad (6.8)$$

and then show that the θ_i's do not simply converge to each other, but they converge to a constant.

At this point, you may want to pause for a moment and compare the flocking problem under consideration to the one in Chap. 3. The model (6.7) is a rotational double-integrator, while the model (3.1) in Chap. 3 is a single rotational integrator. Since the control input here is an acceleration rather than a speed, we cannot directly apply the Kuramoto-inspired control law (3.5) to solve the flocking problem. Not surprisingly, though, the solution we are about to present is a straightforward adaptation of the Kuramoto-inspired control law.

Inspired by [11, 12], we begin our control design by a mechanical analogy. Consider the point particle in Fig. 6.12. The particle has mass J_i and is constrained to move around a circular track of radius 1 metre. A force τ_i is applied to the particle in the direction tangent to the circle. The motion of such a particle is described by (6.7). Now imagine a collection of such masses moving on a common unit circle with the same visibility graph $\mathcal{G}$ as the flying robots. Solving the flocking problem for the flying robots is equivalent to determining admissible forces τ_i synchronizing the particles on the unit circle.

Consider the three particles i, k, l in Fig. 6.13, with particles k, l neighbours of particle i. If we connect the particles by massless springs and add a damping force to each mass, it is reasonable to expect that the particles will synchronize provided their initial conditions are not too far apart and there are enough springs (this latter requirement will translate to requiring that the visibility graph be connected).

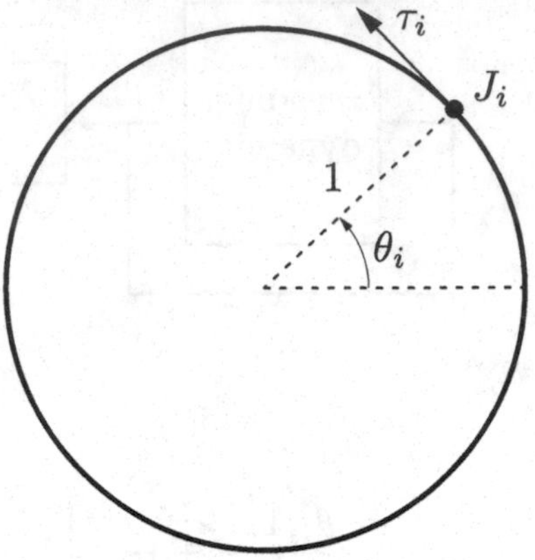

Fig. 6.12 Mechanical
analogy for flocking control
design

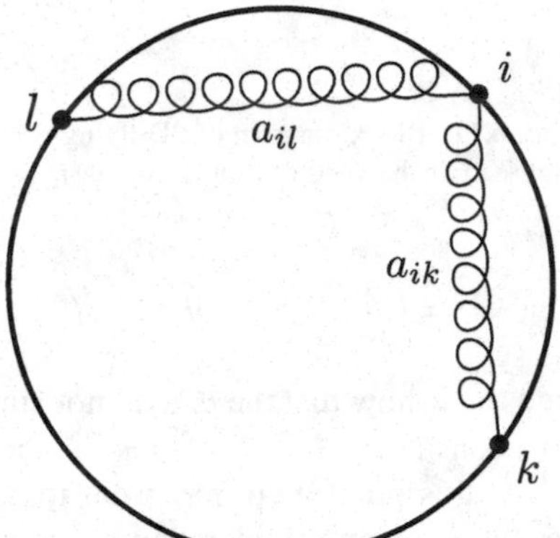

Fig. 6.13 Mechanical
analogy for synchronization
mechanism

What is the control law τ_i corresponding to the setup of Fig. 6.13? The answer is in the next lemma.

Lemma 6.1 *Consider n point particles constrained to move on the unit circle. Suppose the particles are connected by springs as described above, and each particle is subjected to a viscous friction force. The model of the particles is*

$$J_i \ddot{\theta}_i = -b_i \dot{\theta}_i - \sum_{k \in \mathcal{N}_i} a_{ik} \sin(\theta_i - \theta_k), \quad i = 1, \ldots, n. \tag{6.9}$$

Proof In order to avoid the computation of the reaction forces arising from the fact that the particles move on a unit circle we take the Lagrangian approach.

The total kinetic energy of the collection of particles is

$$K = \sum_{l=1}^{n} \frac{1}{2} J_l (\dot{\theta}_l)^2.$$

Letting d_{lk} denote the length of the chord connecting robots l and k and $a_{lk} > 0$ denote the associated spring constant (in what follows, we pick $a_{kl} = a_{lk}$), we find that the total potential energy is the sum of the potential energies of the springs:

$$U = \sum_{l=1}^{n} \sum_{k \in \mathcal{N}_l, k < l} \frac{1}{2} a_{lk} d_{lk}^2.$$

The condition $k < l$ in the inner sum guarantees that the potential energy of the spring connecting robots l and k is counted only once. It can be shown that $d_{lk}^2 = 2 - 2\cos(\theta_l - \theta_k)$. Substituting this expression into U, we obtain the Lagrangian

$$\mathcal{L} = K - U = \sum_{l=1}^{n} \left[\frac{1}{2} J_l (\dot{\theta}_l)^2 - \sum_{k \in \mathcal{N}_l, k < l} a_{lk} (1 - \cos(\theta_l - \theta_k)) \right].$$

This is the Lagrangian of the system of particles subject to the spring forces.

Using the Euler–Lagrange equation

$$\frac{d}{dt} \frac{\partial \mathcal{L}}{\partial \dot{\theta}_i} - \frac{\partial \mathcal{L}}{\partial \theta_i} = 0, \quad i = 1, \ldots, n,$$

we get (the steps are omitted)

$$J_i \ddot{\theta}_i + \sum_{k \in \mathcal{N}_i} a_{ik} \sin(\theta_i - \theta_k) = 0, \quad i = 1, \ldots, n.$$

Adding viscous friction to each particle, we obtain (6.9). □

Comparing (6.7) and (6.9) we deduce the control law for robot i:

$$\tau_i = -b_i \dot{\theta}_i - \sum_{k \in \mathcal{N}_i} a_{ik} \sin(\theta_i - \theta_k). \tag{6.10}$$

This control law is admissible. Does it solve the flocking problem? The answer is found in the next theorem.

Theorem 6.1 *If the visibility graph is constant, undirected, and connected, then for any $a_{ik} = a_{ki} > 0$, $b_i > 0$, $i, k = 1, \ldots, n$, the control law (6.10) makes the flocking manifold $\mathcal{F}$ in (6.8) locally asymptotically stable for (6.7) and solves the flocking problem.*

Proof The total energy of the particles,

$$E = K + U = \sum_{i=1}^{n} \left[\frac{1}{2} J_i (\dot{\theta}_i)^2 + \sum_{k \in \mathcal{N}_i, k < i} a_{ik} (1 - \cos(\theta_i - \theta_k)) \right],$$

is a nonnegative function. We claim that the level set $\{E = 0\}$ coincides with the flocking manifold $\mathcal{F}$ in (6.8). Indeed, since the energy is a sum of nonnegative terms, $E = 0$ if and only if

$$\dot{\theta}_i = 0 \text{ and } 1 - \cos(\theta_i - \theta_k) = 0, \quad i = 1, \ldots, n, \quad k \in \mathcal{N}_i, \quad k < i.$$

Thus $\theta_i = \theta_k$ for all $i \in \{1, \ldots, n\}$ and all $k \in \mathcal{N}_i, k < i$. Since the visibility graph is connected, this latter condition is equivalent to $\theta_1 = \cdots = \theta_n$. Thus $\{E = 0\} = \mathcal{F}$ and the claim is proved.

The time derivative of E along solutions is given by

$$\dot{E} = - \sum_{i=1}^{n} b_i (\dot{\theta}_i)^2 \leq 0.$$

Thus the energy is nonincreasing along solutions, which implies that the flocking manifold $\mathcal{F}$ is stable.

Let $\underline{a} := \min a_{ik}$, and for a given $\gamma \in (0, \pi/2)$, define the energy sublevel set

$$E_\gamma := \{(\theta_1, \ldots, \theta_n, \dot{\theta}_1, \ldots, \dot{\theta}_n) : E(\theta_1, \ldots, \theta_n, \dot{\theta}_1, \ldots, \dot{\theta}_n) < \underline{a}(1 - \cos \gamma)\}.$$

Since E is nonincreasing along solutions, E_γ is positively invariant for any γ, that is, all solutions of (6.9) initialized in E_γ remain there for all $t \geq 0$. Moreover, solutions in E_γ satisfy $(1 - \cos(\theta_i(t) - \theta_k(t))) < (1 - \cos(\gamma))$, or

$$(\forall t \geq 0)(\forall i \in \{1, \ldots, n\})(\forall k \in \mathcal{N}_i) \, |\theta_i(t) - \theta_k(t)| < \gamma. \tag{6.11}$$

To visualize E_γ note that, in it, neighbouring particles lie on an arc of the unit circle whose central angle is 2γ —see Fig. 6.14. Moreover, their speeds have a bound that depends on γ. We will see that if the arc in question and the speeds are sufficiently small (i.e., γ is small), then the particles synchronize.

Fig. 6.14 Interpretation of set E_γ

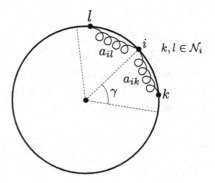

By the LaSalle invariance principle,[3] solutions of (6.9) initialized in E_γ converge to the largest invariant subset of $E_\gamma \cap \{\dot{\theta}_i = 0, i = 1, \ldots, n\}$. We will show that for sufficiently small γ, this set is contained in the flocking manifold $\mathcal{F}$. To this end, note that if $\dot{\theta}_i(t) \equiv 0$, then $\ddot{\theta}_i(t) \equiv 0$, implying that $\sum_{k \in \mathcal{N}_i} a_{ik} \sin(\theta_i - \theta_k) \equiv 0$. Define a Laplacian matrix L as follows:

$$
\begin{aligned}
L_{ik} &:= -a_{ik}, & k \in \mathcal{N}_i, k \neq i, \\
L_{ik} &:= 0, & k \notin \mathcal{N}_i, k \neq i, \\
L_{ii} &:= \sum_{k \in \mathcal{N}_i} a_{ik}.
\end{aligned}
$$

Then L is the Laplacian of the visibility graph $\mathcal{G}$ with positive weights a_{ik} on the graph's edges. Since[4] $\mathcal{G}$ is connected, L has rank $n - 1$ and $\ker L$ is spanned by $\mathbf{1}$.

Let $\theta = (\theta_1, \ldots, \theta_n)$ and define the vector function $r(\theta)$ as

$$
r_i(\theta) := \sum_{k \in \mathcal{N}_i} a_{ik} \left(\sin(\theta_i - \theta_k) - (\theta_i - \theta_k) \right), \quad i = 1, \ldots, n.
$$

Then the condition $\sum_{k \in \mathcal{N}_i} a_{ik} \sin(\theta_i - \theta_k) \equiv 0, i = 1, \ldots, n$, may be expressed as

$$
L\theta + r(\theta) = 0.
$$

The above identity can hold only if $\|L\theta\| = \|r(\theta)\|$. It can be shown that

$$
\lim_{L\theta \to 0} \frac{r_i(\theta)}{\|L\theta\|} = 0.
$$

Therefore, for sufficiently small γ, the property (6.11) implies that

$$
\|r(\theta)\| \leq \|L\theta\|/2.
$$

For such small γ, the unique solution of

$$
L\theta + r(\theta) = 0
$$

is $L\theta = 0$, or $\theta_1 = \cdots = \theta_n$.

We have thus shown that the largest invariant subset of $E_\gamma \cap \{\dot{\theta}_i = 0, i = 1, \ldots, n\}$ is contained in $\mathcal{F}$. By the LaSalle invariance principle, the flocking manifold $\mathcal{F}$ is locally asymptotically stable, and the set E_γ is contained in its domain of attraction.

---

[3] The LaSalle invariance principle requires solutions to be bounded. The speeds $\dot{\theta}_i$ are bounded because of the damping term $-b_i \dot{\theta}_i$ in (6.9). As for the angles θ_i, we view them as points of a unit circle, a compact set.

[4] This result is analogous to Theorem 4.2, which covers the case of unit weights, $a_{lk} = 1$ for all l, k.

We are left to show that θ_i converges to a constant. This part requires a little more work, and it involves the centre manifold theorem. We omit the argument, but you may look at the proof of Theorem 4 in [20] to get the idea.   □

6.4 Flocking of 3D Flying Robots

In this section we generalize the flocking control law to 3D flying robots. The generalization relies on the same mechanical analogy but now, instead of imagining point particles on the unit circle, we imagine rigid bodies on the unit sphere.

Consider a collection of n robots, each one modelled by (6.3). As before, the flocking problem[5] is to find an admissible control law making the velocity vectors $\dot{x}_i$ converge to a common constant vector v_{ss} dependent on the initial conditions. We assume, as before, that the visibility graph is constant and undirected, and that each robot is affected by a drag force pointing opposite to its velocity vector. We now also assume that the robots have identical masses m, so that their translational motions are modelled by

$$m\ddot{x}_i = mge_3 - u_i R_i e_3 - b\dot{x}_i.$$

As in the 2D case, if we set $u_1 = \cdots = u_n = \bar{u} > 0$ and we make $R_i e_3$ converge to a common constant vector, then the velocities $\dot{x}_i$ converge to a common constant vector as well, and flocking is achieved. The problem has thus been reduced to the synchronization of the vectors $R_i e_3, i = 1, \ldots, n$, which we will refer to as the **thrust axes** of the robots. For the design of flocking controllers we focus our attention on the rotational dynamics

$$\dot{R}_i = R_i S(\omega_i)$$
$$J_i \dot{\omega}_i + \omega_i \times (J_i \omega_i) = \tau_i, \quad i = 1, \ldots, n. \tag{6.12}$$

Before continuing our development, we introduce some useful notation, illustrated in Fig. 6.15. We let $q_i := R_i e_3$ denote the thrust axis of robot i represented in the common inertial frame $\mathcal{I}$, and by $R_j^i := R_i^T R_j$ the relative orientation of robot j with respect to robot i. Operationally, R_j^i transforms a vector in frame j to its representation in frame i. We let $q_j^i := R_j^i e_3$ denote the representation of robot j's thrust axis in the local frame of robot i. For flocking we would like to have $q_j^i = e_3$ for all i, j. Finally, we denote by φ_j^i the angle between the third axis of frame $\mathcal{B}_i$ and q_j^i. Flocking corresponds to the condition $\varphi_j^i = 0$ for all i, j.

With the notation just introduced, we have three equivalent definitions of the flocking manifold:

---

[5]In our formulation of the flocking problem, nothing prevents the robots from crashing to the ground. A more meaningful problem statement would require v_{ss} to be parallel to the ground, but this problem is to date open and significantly harder than the one considered in this section.

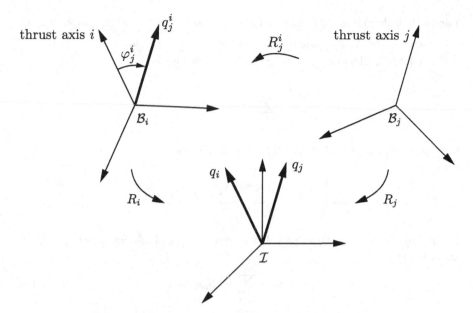

Fig. 6.15 Coordinate representations of the thrust axes of robots i and j

$$
\begin{aligned}
\mathcal{F} := &\{(R_1, \ldots, R_n, \omega_1, \ldots, \omega_n) : q_i = q_j, \ \omega_i = 0, \ i, j = 1, \ldots, n\} \\
= &\{(R_1, \ldots, R_n, \omega_1, \ldots, \omega_n) : q_j^i = e_3, \ \omega_i = 0, \ i, j = 1, \ldots, n\} \\
= &\{(R_1, \ldots, R_n, \omega_1, \ldots, \omega_n) : \varphi_j^i = 0, \ \omega_i = 0, \ i, j = 1, \ldots, n\}.
\end{aligned}
$$

$$(6.13)$$

Thus, on $\mathcal{F}$, the robots have zero angular velocity and identical thrust axes. This latter requirement is expressed in three equivalent ways in (6.13). The first identity expresses it in the common inertial frame $\mathcal{I}$, while the second and third identities express it in the robots' body frames.

In the context of 3D flying robots, an admissible control law for the rotational dynamics of robot i is one that depends only on the quantities

$$
R_k^i, \quad \omega_i, \quad k \in \mathcal{N}_i
$$

and is a locally Lipschitz function.

Since the vectors q_i have unit length, the flocking problem can be regarded as one of synchronization on the unit sphere. In analogy with the 2D case, we imagine that the thrust axis q_i is the position of the centre of mass of a rigid body on the unit sphere. We regard the control τ_i as a force applied at the centre of mass of the body in a direction tangent to the sphere. Once again, we aim to synchronize these imaginary rigid bodies by connecting them by springs and adding damping to each of them, with the convention that a spring with constant $a_{ik} > 0$ interconnects bodies i and k

if and only if there is an edge between nodes i and k of the visibility graph. It is an easy matter to check that the length of this spring is $\sqrt{2(1 - \langle q_k^i, e_3 \rangle)}$.

The total potential energy of the collection of springs is

$$U = \sum_{i=1}^{n} \sum_{k \in \mathcal{N}_i, k < i} a_{ik}(1 - \langle q_k^i, e_3 \rangle).$$

The total energy of the robots connected by springs is

$$E = \sum_{i=1}^{n} \frac{1}{2} \omega_i^T J_i \omega_i + \sum_{i=1}^{n} \sum_{k \in \mathcal{N}_i, k < i} a_{ik}(1 - \langle q_k^i, e_3 \rangle).$$

Inspired by the proof of Theorem 6.1, we define a control law for system (6.12) by requiring that

$$\dot{E} = -\sum_{i=1}^{n} \omega_i^T B_i \omega_i, \qquad (6.14)$$

where B_i is a symmetric positive definite matrix. By imposing this identity, we get the control law

$$\tau_i = -B_i \omega_i - \sum_{k \in \mathcal{N}_i} a_{ik}(q_k^i \times e_3), \qquad (6.15)$$

which can also be expressed as

$$\tau_i = -B_i \omega_i - R_i^T \sum_{k \in \mathcal{N}_i} a_{ik}(q_k \times q_i).$$

We postpone the verification that this control law gives identity (6.14) to the proof of Theorem 6.2. Note that this control law is admissible, as it relies only on R_i^k, $k \in \mathcal{N}_i$ and ω_i.

At this point you may want to reflect on the similarity between (6.15) and its 2D counterpart (6.10). As an exercise, show that (6.10) is a special case of (6.13), in the following sense. Suppose the third body frame axis of each robot is parallel to the third axis of the inertial frame $\mathcal{I}$, i.e., $R_i e_3 = e_3$. Thus the rotation matrix of robot i has the form

$$R_i = \begin{bmatrix} \cos(\theta_i) & -\sin(\theta_i) & 0 \\ \sin(\theta_i) & \cos(\theta_i) & 0 \\ 0 & 0 & 1 \end{bmatrix}.$$

Suppose further that the thrust axis is parallel to the first body axis rather than the third. Thus $q_j^i = R_j^i e_1$. Finally, let B_i be a diagonal matrix with $(B_i)_{33} = b_i$. Then τ_i in (6.15) has the form $\tau_i = (0, 0, (\tau_i)_3)$, where $(\tau_i)_3$ coincides with (6.10).

Theorem 6.2 *If the visibility graph is constant, undirected, and connected, then for every* $a_{ik} = a_{ki} > 0$ *and every symmetric positive definite matrix* B_i, $i, k = 1, \ldots, n$, *the control law (6.15) makes the flocking manifold* $\mathcal{F}$ *in (6.13) locally asymptotically stable for (6.12) and solves the flocking problem.*

Proof The energy E is nonnegative. It is zero if and only if $\omega_i = 0$ and $\langle q_k^i, e_3 \rangle = 1$ for all $i = 1, \ldots, n$ and $j \in \mathcal{N}_i$. Since q_k^i is a unit vector, the latter condition is equivalent to $q_k^i = e_3$. Since the visibility graph is connected, this identity is true for all $i, k = 1, \ldots, n$, and therefore the level set $\{E = 0\}$ coincides with the flocking manifold $\mathcal{F}$ in (6.13).

Next we show that (6.14) holds. We have

$$\dot{E} = \sum_{i=1}^{n} \omega_i^T J_i \dot{\omega}_i - \sum_{i=1}^{n} \sum_{k \in \mathcal{N}_i, k < i} a_{ik} \langle \dot{q}_k^i, e_3 \rangle.$$

Using the dynamics (6.12) with control (6.15), we get

$$\dot{E} = -\sum_{i=1}^{n} \omega_i^T B_i \omega_i - \sum_{i=1}^{n} \sum_{k \in \mathcal{N}_i} a_{ik} \omega_i^T (q_k^i \times e_3) - \sum_{i=1}^{n} \sum_{k \in \mathcal{N}_i, k < i} a_{ik} \langle \dot{q}_k^i, e_3 \rangle.$$

Recalling the definition of q_k^i, $q_k^i = R_i^T R_k e_3$, and differentiating this identity with respect to time, we obtain

$$\begin{aligned}
\dot{q}_k^i &= \dot{R}_i^T R_k e_3 + R_i^T \dot{R}_k e_3 \\
&= S(\omega_i)^T R_i^T R_k e_3 + R_i^T R_k S(\omega_k) e_3 \\
&= -S(\omega_i) q_k^i + R_k^i S(\omega_k) e_3.
\end{aligned}$$

In the second identity we used the derivative of a rotation matrix in (6.12). For any rotation matrix R and any vector $v \in \mathbb{R}^3$, we have

$$R S(v) R^T = S(Rv).$$

Using this identity and the linearity of the operator S, we obtain

$$\begin{aligned}
\dot{q}_k^i &= S(-\omega_i + R_k^i \omega_k) q_k^i \\
&= (-\omega_i + R_k^i \omega_k) \times q_k^i \\
&= q_k^i \times (\omega_i - R_k^i \omega_k) \\
&= S(q_k^i)(\omega_i - R_k^i \omega_k).
\end{aligned}$$

We now take the inner product with e_3:

$$\begin{aligned}
\langle \dot{q}_k^i, e_3 \rangle &= (\omega_i - R_k^i \omega_k)^T S(q_k^i)^T e_3 \\
&= -(\omega_i - R_k^i \omega_k)^T S(q_k^i) e_3 \\
&= -\omega_i^T S(q_k^i) e_3 + \omega_k^T (R_k^i)^T S(q_k^i) e_3 \\
&= -\omega_i^T (q_k^i \times e_3) + \omega_k^T S((R_k^i)^T q_k^i)(R_k^i)^T e_3 \\
&= -\omega_i^T (q_k^i \times e_3) + \omega_k^T S(e_3) q_i^k \\
&= -\omega_i^T (q_k^i \times e_3) - \omega_k^T (q_i^k \times e_3).
\end{aligned}$$

Using this result in the third term of the equation for $\dot{E}$, we get

$$\begin{aligned}
\dot{E} = &-\sum_{i=1}^{n} \omega_i^T B_i \omega_i - \sum_{i=1}^{n} \sum_{k \in \mathcal{N}_i} a_{ik} \omega_i^T (q_k^i \times e_3) \\
&+ \sum_{i=1}^{n} \sum_{k \in \mathcal{N}_i, k<i} a_{ik} \omega_i^T (q_k^i \times e_3) \\
&+ \sum_{i=1}^{n} \sum_{k \in \mathcal{N}_i, k<i} a_{ik} \omega_k^T (q_i^k \times e_3).
\end{aligned}$$

Since the graph is undirected and $a_{ik} = a_{ki}$, the last sum can be rewritten as

$$\sum_{k=1}^{n} \sum_{i \in \mathcal{N}_k, i>k} a_{ki} \omega_k^T (q_i^k \times e_3).$$

Now swap the indices i and k. Then the last three sums in the above expression for $\dot{E}$ evaluate to zero, and

$$\dot{E} = -\sum_{i=1}^{n} \omega_i^T B_i \omega_i,$$

as required.

Let $\underline{a} := \min a_{ik}$, and for a given $\gamma \in (0, \pi/2)$, define the energy sublevel set $E_\gamma := \{(R_1, \ldots, R_n, \omega_1, \ldots, \omega_n) : E(R_1, \ldots, R_n, \omega_1, \ldots, \omega_n) < \underline{a}(1 - \cos\gamma)\}$. As in the proof of Theorem 6.1, E_γ is positively invariant and all solutions initialized in it satisfy $\langle q_k^i(t), e_3 \rangle = \cos\varphi_k^i(t) > \cos\gamma$, or

$$(\forall t \geq 0)(\forall i \in \{1, \ldots, n\})(\forall k \in \mathcal{N}_i) \, |\varphi_k^i(t)| < \gamma.$$

Since the visibility graph $\mathcal{G}$ is connected, we may pick γ small enough that

$$\left((\forall i \in \{1, \ldots, n\})(\forall k \in \mathcal{N}_i)\ |\varphi_k^i| < \gamma\right) \implies \max_{i,k \in \{1,\ldots,n\}} \{|\varphi_k^i|\} < \frac{\pi}{2}.$$

Thus all solutions initialized in E_γ have the property that their thrust axes $q_i(t)$ lie on a common half plane for all time. More precisely, for all initial conditions in E_γ, the solutions of (6.12) with control (6.15) satisfy

$$(\forall t \geq 0)(\exists v \in \mathbb{R}^n, v \neq 0)(\forall i \in \{1, \ldots, n\})\ \langle q_i(t), v \rangle > 0. \qquad (6.16)$$

By (6.14) the energy E is nonincreasing along solutions. Since its zero level set is the flocking manifold $\mathcal{F}$, $\mathcal{F}$ is a stable set. By the LaSalle invariance principle,[6] all solutions of the closed-loop system initialized in E_γ converge to the largest invariant set contained in $E_\gamma \cap \{\omega_i = 0, i = 1, \ldots, n\}$. We will show that this set is contained in the flocking manifold.

If $\omega_i(t) \equiv 0$, then $\dot{\omega}_i(t) \equiv 0$ and so $J_i \dot{\omega}_i(t) \equiv 0$, implying that the control τ_i in (6.15) is identically zero. Thus, we have

$$(\forall i \in \{1, \ldots, n\}) \sum_{k \in \mathcal{N}_i} a_{ik}(q_k^i(t) \times e_3) \equiv 0.$$

Premultiplying this identity by $R_i(t)$ we get

$$(\forall i \in \{1, \ldots, n\}) \sum_{k \in \mathcal{N}_i} a_{ik}(q_k(t) \times q_i(t)) \equiv 0,$$

or

$$(\forall i \in \{1, \ldots, n\})\ q_i(t) \times \left(\sum_{k \in \mathcal{N}_i} a_{ik}q_k(t)\right) \equiv 0.$$

Thus $q_i(t)$ and $\sum_{k \in \mathcal{N}_i} a_{ik}q_k(t)$ are parallel, i.e., there exists a scalar $\lambda_i(t)$ such that

$$q_i(t) = \lambda_i(t) \sum_{k \in \mathcal{N}_i} a_{ik}q_k(t). \qquad (6.17)$$

We claim that $\lambda_i(t) > 0$. Indeed, taking the inner product of both sides of (6.17) with the vector v in (6.16) we obtain

$$\langle q_i(t), v \rangle = \lambda_i(t) \sum_{k \in \mathcal{N}_i} a_{ik}\langle q_k(t), v \rangle.$$

---

[6]As in the proof of Theorem 6.1, we may apply the LaSalle invariance principle because all solutions of (6.12) with control (6.15) are bounded. The boundedness of ω_i follows from the presence of the dissipation term $-B_i\omega_i$ in the τ_i. The matrices R_i have unit norm columns so they are bounded as well.

By property (6.16), the left-hand side is positive. The sum in the right-hand side is also positive because $a_{ik} > 0$ and each inner product is positive. Thus it must hold that $\lambda_i(t) > 0$, as claimed. Since q_i has unit norm we conclude that

$$\lambda_i(t) = \left(\left\| \sum_{k \in \mathcal{N}_i} a_{ik} q_k(t) \right\| \right)^{-1}. \tag{6.18}$$

From now on, we drop the time dependence on all variables. Consider the following weighted sum of the q_i's, and use identity (6.17):

$$\sum_{i=1}^{n} \left(\sum_{k \in \mathcal{N}_i} a_{ik} \right) q_i = \sum_{i=1}^{n} \lambda_i \left(\sum_{k \in \mathcal{N}_i} a_{ik} \right) \sum_{l \in \mathcal{N}_i} a_{il} q_l.$$

The right-hand side of this identity is a linear combination of $\{q_1, \ldots, q_n\}$. A generic term q_j appears in the right-hand side when $l = j$ and $j \in \mathcal{N}_i$, or since the visibility graph is undirected, $l = j, i \in \mathcal{N}_j$. Thus the above identity can be rewritten as

$$\sum_{i=1}^{n} \left(\sum_{k \in \mathcal{N}_i} a_{ik} \right) q_i = \sum_{j=1}^{n} \mu_j q_j,$$

with

$$\mu_j = \sum_{i \in \mathcal{N}_j} \lambda_i \left(\sum_{k \in \mathcal{N}_i} a_{ik} \right) a_{ij}.$$

We thus have

$$\sum_{i=1}^{n} \left[\left(\sum_{k \in \mathcal{N}_i} a_{ik} \right) - \mu_i \right] q_i = 0. \tag{6.19}$$

Consider the definition of λ_i in (6.18). Since the q_i's have unit norm and $a_{ik} > 0$, by the triangle inequality we have

$$\lambda_i \geq \left(\sum_{k \in \mathcal{N}_i} a_{ik} \right)^{-1}.$$

This lower bound on λ_i gives a lower bound on μ_j,

$$\mu_j \geq \sum_{i \in \mathcal{N}_j} a_{ij}.$$

Thus the coefficient of q_i in (6.19) is upper bounded as follows

$$\left(\sum_{k \in \mathcal{N}_i} a_{ik}\right) - \mu_i \le \sum_{k \in \mathcal{N}_i} a_{ik} - \sum_{k \in \mathcal{N}_i} a_{ki} = 0.$$

In conclusion, the left-hand side of (6.19) is a linear combination of all q_i's with nonpositive coefficients. By property (6.16), identity (6.19) can hold only if

$$\left(\sum_{k \in \mathcal{N}_i} a_{ik}\right) - \mu_i = 0 \quad \text{for all } i = 1, \dots, n,$$

which implies that $\lambda_i = \left(\sum_{k \in \mathcal{N}_i} a_{ik}\right)^{-1}, i = 1, \dots, n,$ or

$$\left\| \sum_{k \in \mathcal{N}_i} a_{ik} q_k \right\| = \sum_{k \in \mathcal{N}_i} a_{ik} \quad \text{for all } i = 1, \dots, n.$$

This identity implies that, for each $i \in \{1, \dots, n\}$, the vectors $\{q_k : k \in \mathcal{N}_i\}$ are parallel to each other. By (6.17), the vectors $\{q_i, q_k : k \in \mathcal{N}_i\}$ are also parallel. We have thus established that if robots i and k are neighbours in the visibility graph, then q_i and q_k are parallel. Since the visibility graph is connected, this property implies that $q_1, \dots, q_n$ are parallel. Finally, the fact that the q_i's have unit norm and lie on a common half-plane (property (6.16)) implies that $q_1 = \cdots = q_n$.

We have thus shown that the largest invariant subset of $E_\gamma \cap \{\omega_i = 0, i = 1, \dots, n\}$ is the flocking manifold $\mathcal{F}$. By the LaSalle invariance principle, all solutions originating in E_γ converge to $\mathcal{F}$, and therefore $\mathcal{F}$ is asymptotically stable. The proof that the q_i's converge to a constant is omitted.   □

6.5 Rendezvous of 3D Flying Robots

The rendezvous problem is the gateway to more complex coordination problems such as the control of formations. The idea, as in Chap. 4, is to get n identical flying robots to convene using only feedback from onboard sensors. More precisely, consider n robots modelled by

$$m\ddot{x}_i = mge_3 - u_i R_i e_3$$
$$\dot{R}_i = R_i S(\omega_i) \tag{6.20}$$
$$J\dot{\omega}_i = -\omega_i \times (J\omega_i) + \tau_i,$$

with state $\chi := (\chi_1, \dots, \chi_n)$, $\chi_i := (x_i, \dot{x}_i, R_i, \omega_i)$. We define the **rendezvous manifold**

$$\mathcal{R} = \{\chi : x_1 = \cdots = x_n, \dot{x}_1 = \cdots = \dot{x}_n\}.$$

On this set, the centres of mass and velocities of all robots coincide, but the robots are not necessarily stationary. No constraint is placed on the robots' orientations and angular velocities.

As before, an admissible control for robot i is a locally Lipschitz function of the quantities

$$R_i^T(x_k - x_i), \quad R_i^T(\dot{x}_k - \dot{x}_i), \quad R_i^T R_k, \quad \omega_i,$$

for each $k \in \mathcal{N}_i$.

We assume that the visibility graph is constant. The **rendezvous problem** for the robots in (6.20) is to find, if they exist, admissible controls (u_i, τ_i) so that there exists a subset of the rendezvous manifold that is locally asymptotically stable.

Allowing the stabilization of a subset of $\mathcal{R}$, rather than the entire $\mathcal{R}$, makes the problem less demanding and allows one to take different lines of attack for its solution. For instance, one may attempt to locally asymptotically stabilize the subset of $\mathcal{R}$ on which the thrust axes coincide,

$$\{\chi \in \mathcal{R} : R_1 e_3 = \cdots = R_n e_3\},$$

in which case the requirement on the initial conditions would be that the robots' initial positions, velocities, and thrust axes be close to each other. Or one may even attempt to fully synchronize the rotation matrices R_i. In another formulation, one may require the robots to stop moving, in which case the subset of $\mathcal{R}$ of interest would be

$$\{\chi \in \mathcal{R} : \dot{x}_i = \cdots = \dot{x}_n = 0\}.$$

Combinations of the above formulations are of course possible, and they are all special instances of the general problem statement above. One may also formulate the global version of the rendezvous problem, or other variations along these lines (almost-global, semi-global, practical, and so on).

Even in our weak formulation, the rendezvous problem for flying robots is to date open. What makes the problem particularly hard is the requirement that robot i must be able to compute its own control (u_i, τ_i) without knowing its absolute position x_i and orientation R_i in the common inertial frame $\mathcal{I}$.

Series Editors' Biography

Tamer Başar is with the University of Illinois at Urbana-Champaign, where he holds the academic positions of Swanlund Endowed Chair, Center for Advanced Study Professor of Electrical and Computer Engineering, Research Professor at the Coordinated Science Laboratory, and Research Professor at the Information Trust Institute. He received the B.S.E.E. degree from Robert College, Istanbul, and the M.S., M.Phil, and Ph.D. degrees from Yale University. He has published extensively in systems, control, communications, and dynamic games, and has current research interests that address fundamental issues in these areas along with applications such as formation in adversarial environments, network security, resilience in cyber-physical systems, and pricing in networks.

In addition to his editorial involvement with these Briefs, Basar is also the Editor-in-Chief of Automatica, Editor of two Birkhäuser Series on Systems & Control and Static & Dynamic Game Theory, the Managing Editor of the Annals of the International Society of Dynamic Games (ISDG), and member of editorial and advisory boards of several international journals in control, wireless networks, and applied mathematics. He has received several awards and recognitions over the years, among which are the Medal of Science of Turkey (1993); Bode Lecture Prize (2004) of IEEE CSS; Quazza Medal (2005) of IFAC; Bellman Control Heritage Award (2006) of AACC; and Isaacs Award (2010) of ISDG. He is a member of the US National Academy of Engineering, Fellow of IEEE and IFAC, Council Member of IFAC (2011–2014), a past president of CSS, the founding president of ISDG, and president of AACC (2010–2011).

Antonio Bicchi is Professor of Automatic Control and Robotics at the University of Pisa. He graduated from the University of Bologna in 1988 and was a postdoc scholar at M.I.T. A.I. Lab between 1988 and 1990. His main research interests are in:

- dynamics, kinematics and control of complex mechanical systems, including robots, autonomous vehicles, and automotive systems;
- haptics and dextrous manipulation; and theory and control of nonlinear systems, in particular hybrid (logic/dynamic, symbol/signal) systems.

© The Author(s) 2016
B.A. Francis and M. Maggiore, *Flocking and Rendezvous in Distributed Robotics*, SpringerBriefs in Control, Automation and Robotics, DOI 10.1007/978-3-319-24729-8

- theory and control of nonlinear systems, in particular hybrid (logic/dynamic, symbol/signal) systems.

He has published more than 300 papers in international journals, books, and refereed conferences.

Professor Bicchi currently serves as the Director of the Interdepartmental Research Center "E. Piaggio" of the University of Pisa, and President of the Italian Association or Researchers in Automatic Control. He has served as Editor in Chief of the Conference Editorial Board for the IEEE Robotics and Automation Society (RAS), and as Vice President of IEEE RAS, Distinguished Lecturer, and Editor for several scientific journals including the *International Journal of Robotics Research, the IEEE Transactions on Robotics and Automation, and IEEE RAS Magazine*. He has organized and co-chaired the first WorldHaptics Conference (2005), and Hybrid Systems: Computation and Control (2007). He is the recipient of several best paper awards at various conferences, and of an Advanced Grant from the European Research Council. Antonio Bicchi has been an IEEE Fellow since 2005.

Miroslav Krstic holds the Daniel L. Alspach chair and is the founding director of the Cymer Center for Control Systems and Dynamics at University of California, San Diego. He is a recipient of the PECASE, NSF Career, and ONR Young Investigator Awards, as well as the Axelby and Schuck Paper Prizes. Professor Krstic was the first recipient of the UCSD Research Award in the area of engineering and has held the Russell Severance Springer Distinguished Visiting Professorship at UC Berkeley and the Harold W. Sorenson Distinguished Professorship at UCSD. He is a Fellow of IEEE and IFAC. Professor Krstic serves as Senior Editor for *Automatica and IEEE Transactions on Automatic Control* and as Editor for the Springer series *Communications and Control Engineering*. He has served as Vice President for Technical Activities of the IEEE Control Systems Society. Krstic has co-authored eight books on adaptive, nonlinear, and stochastic control, extremum seeking, control of PDE systems including turbulent flows and control of delay systems.

Appendix
On the Literature

In this concluding section we discuss some relevant literature and specific references. Flocking and rendezvous in distributed robotics have precursors and analogies in other scientific domains, and that is something we shall look at now.

Robot Models

The literature on distributed robotics has several threads. One is in the context of computer science and robots are modeled as formal computing devices, somewhat similar to Turing machines, and may sometimes be called "processes". A good example of such a paper is [42] by Suzuki and Yamashita. It begins with an interesting puzzle by way of motivation: Imagine an elementary school teacher and her class of kids in the school playground. Suppose she wants all the kids themselves to form a circle. Is it possible for the teacher to give an instruction simultaneously to all the kids so that they will proceed to form a circle?

In contrast, in this book robot models are differential equations derived from physics.

Flocking

The word "biomimetic" refers to human-made processes, substances, devices, or systems that imitate nature. An archetypal biomimetic problem is to design and build a group of robots that have the ability to fly and to flock while flying. In 1987 Craig Reynolds introduced a model and wrote a program called **boids** (android birds) that simulates a flock of birds in flight; they fly as a flock, with a common average heading, and they avoid colliding with each other. Each boid has a local control

B.A. Francis and M. Maggiore, *Flocking and Rendezvous in Distributed Robotics*, SpringerBriefs in Control, Automation and Robotics, DOI 10.1007/978-3-319-24729-8

strategy—there is no leader broadcasting instructions—yet a desirable overall group behaviour is achieved. The local strategy of each boid has three components:

1. Separation: steer to avoid crowding;
2. Alignment: steer towards the average heading of neighbours;
3. Cohesion: steer towards the average position of neighbours.

Following Reynolds is the paper [46] by Vicsek et al. that appeared in 1995. They propose a simple discrete-time model of n autonomous agents i.e., points or particles all moving in the plane with the same speed but with different headings. Each agent's heading is updated using a local rule based on the average of its own heading together with the headings of its neighbours. Agent i's neighbours at time t are those agents that are either in or on a circle of pre-specified radius centred at agent i's current position. The paper provides a variety of interesting simulation results that demonstrate that the nearest-neighbour rule can cause all agents eventually to move in the same direction despite the absence of centralized coordination and despite the fact that each agent's set of nearest neighbours changes with time as the system evolves.

Following Vicsek et al. is the paper [16] by Jadbabaie et al., considering the same motion control law. Jadbabaie et al. prove that flocking occurs provided there is a sufficient amount of connectivity over time of the visibility graph. But there are limitations to the setup and results, as shown in Sect. 3.4.

In [18] Justh and Krishnaprasad solved the simplest flocking problem: two unicycles always in sight of each other. Their controller is admissible in the sense we use in this book. The main result states that if the unicycles are initialized at different locations in the plane and they are not initially heading in opposite directions, then they flock without colliding.

The solution to the unicycle flocking problem presented in Chap. 3, inspired by the Kuramoto model of coupled oscillators [21], has been derived by a number of researchers in the literature. In particular, Sepulchre et al. [40] (see also Scardovi et al. [39]) state a special case of Theorem 3.1 in which the visibility graph is complete (i.e., each robot can see all other robots). Jadbabaie et al. [17] state a result analogous to Theorem 3.1. Lin et al. [26] present a more general result. They show that the Kuramoto-inspired flocking controller solves the flocking problem when the visibility graph is directed, time-varying, and enjoys a suitable connectivity property.

In articles [16, 46] the update law of the heading angles operates in discrete time. One may take issue with this framework. Motion in nature is normally understood to be a continuous-time phenomenon. Permitting two discrete-time bird models in the same system presupposes temporal coordination of sampling; that is, the two clocks internal to the birds must be synchronized by an uber-clock. Allowing such a thing is inconsistent with the objective of studying autonomous agents.

Agreement and Consensus

A group of robots rendezvousing, a problem in robotics, is analogous to a group of agents arriving at a consensus about the value of a number θ, a problem in decision theory. In 1974 DeGroot formulated a problem in the latter subject. His paper, Forming a Consensus [8], begins in this way: Consider a group of n agents who must act together as a team or committee, and suppose that each of these agents can specify its own subjective probability distribution for the value of θ. DeGroot's paper offers a model that describes how the group might reach a consensus and form a common subjective probability distribution for θ simply by revealing their individual distributions to each other and pooling their opinions.

Starting from DeGroot, Tsitsiklis and co-workers have studied distributed decision-making in a large number of papers (see [4, 45], and other papers on his website). In particular, in [4] is reviewed the "agreement algorithm", briefly defined as follows. For each of n agents, $x_i(t)$ denotes the value held by agent i at time t about the value of a parameter θ. The time variable t is taken to be a nonnegative integer. The n held values define a vector $x(t)$, namely, the vector with components $x_1(t), \ldots, x_n(t)$. The update equation for the vector $x(t)$ is taken to be

$$x(k+1) = A(k)x(k), \tag{1}$$

where $A(k)$ is a stochastic matrix, i.e., for every $k \geq 0$, the elements of $A(k)$ are non-negative and its rows sum to 1. Thus, the value of θ held by agent i at time $k+1$ equals the dot product of row i of $A(k)$ with the vector $x(k)$. Consensus is said to occur if there exists a number ϕ such that $x(k)$ converges to $\phi\mathbf{1}$ as $k \to \infty$, where $\mathbf{1}$ denotes the vector of 1's. Whether or not consensus occurs depends on $x(0)$, the initial values, and the matrix function $A(k)$. From a mathematical viewpoint, this agreement problem is equivalent to the flocking problem investigated by Jadbabaie et al. [16], and the conditions for consensus presented in [45] are analogous to the ones in [16] under similar assumptions.

Pursuit

Every motion-control law for a robot outfitted with a camera has to be based on pursuit, for if a robot is "here" and it needs to be "there", and if its only sensor is a camera, then it has to pursue something to get "there". In this way, cyclic pursuit is a strategy for rendezvous.

Mathematical pursuit has a long history that can be traced back to work by the French mathematician Pierre Bouguer [1698–1758]. Much more recently, Martin Gardner wrote a popular mathematics column in the *Scientific American* magazine. Here is his puzzle called *The Four Bugs Problem*, published in 1965: Four bugs, denoted A, B, C, and D, start at the corners of a square. Starting at $t = 0$, A pursues

B at unit speed, B pursues C at unit speed, C pursues D at unit speed, and D pursues A at unit speed. The question is whether or not the bugs meet, and if they do, at what time.

The references for cyclic pursuit are [24, 27].

The Rendezvous Problem

There are essentially two different rendezvous problems, one where neighbour sets are fixed for all time (i.e., the visibility graph does not change with time) and the other where neighbours are defined by proximity. In the former case a control law can be constructed by allowing each robot to pursue the centroid of its neighbour set. It is not clear who first proved the equivalence of conditions 3 and 4 in Theorem 4.2. Certainly Ren and Beard did prove it [34] and so did Lin et al. [25]. In the latter case (where neighbours are determined by proximity), the visibility graph usually varies with time. Allowing each robot to pursue the centroid of its neighbour set will not always work, because visibility links may break. A control law that **can** be proved to work is the circumcentre law presented in Chap. 5, invented by Ando et al. [2]. This control law turns out not to be Lipschitz continuous, and so proof that the control law does indeed provide rendezvous is non-trivial. It was first proved in [26].

The solution of the rendezvous problem for unicycles presented in Theorem 4.3 is taken from [25], where this theorem is proved under more general assumptions.

We have presented solutions of the rendezvous problem for kinematic points and unicycles. Other models have been considered in the literature. For double-integrators, the solution of the rendezvous problem was developed by Ren et al. [31, 32]. Tanner et al. [44] proposed distributed control laws, involving potential functions, that achieve rendezvous while avoiding collisions. Moreau [29] considered the general context of nonlinear discrete-time systems. He proved that if each agent moves towards the relative interior of the convex hull of the set of its neighbours, then rendezvous is achieved. In [26], Lin et al. proved the continuous-time counterpart of Moreau's result. These results assume persistent connectivity of the visibility graph, rather than guaranteeing it.

Flying Robots

The model of a thrust-propelled flying robot presented in Chap. 6 is standard and has been used in more or less the same form by a number of researchers (e.g., [15, 22]), sometimes using the quaternion representation of the aircraft's orientation [1], other times using local representations such as Euler angles [7]. Our model ignores drag effects, wind gusts, and related uncertainties.

Although the flocking problem for underactuated flying robots is not discussed in the literature, the solution for the 2D case we have presented in Theorem 6.1 appears in the work of Dörfler and Bullo [11, 12] in the context of transient stability of power networks. The solution for the 3D case presented in Theorem 6.2 is novel. Attitude synchronization is related to flocking: instead of seeking to synchronize the thrust axes of the robots, in attitude synchronization one wants to synchronize their entire orientation. In this context, Nair and Leonard [30] and Sarlette et al. [38] presented solutions similar to ours.

The rendezvous problem for flying robots remains open. A number of researchers have attempted a solution, but in all cases the control laws are not distributed in that they rely on measurements of the absolute positions or absolute orientations of the robots. Recently, Roza et al. [37] have proposed a distributed solution of the rendezvous problem. This solution, however, requires each robot to measure an inertial vector such as gravity in its own body frame. Moreover, robots are required to communicate certain variables to their neighbours.

Discrete-Event Robots

Versions of the rendezvous problem have been studied extensively in computer science (where it is usually called *the gathering problem*). An example is [14]. Each robot is viewed as a point in the plane. The robots have limited visibility: Each can see only the other robots within a fixed radius. Moreover, the robots are modelled as asynchronous discrete-event systems having four possible states: *Wait*, that is, not moving and idle; *Look*, during which the robot senses the relative positions of the other robots within its field of view; *Compute*, during which it computes its next move; and *Move*, during which it moves at some pre-determined speed to its computed destination. There are soft timing assumptions, such as, a robot can be in Wait for only a finite period of time.

The robots have local coordinate frames and these are assumed to have a common orientation, e.g., they may each have a compass as shown in Fig. 1. The paper proposes the following control law, in the form of four if-then rules:

1. If in the Look state a robot sees another robot to its left or vertically above, then it does not move.
2. If a robot sees robots only below on its vertical axis, then it moves down toward the nearest robot.
3. If a robot sees robots only to its right, then it moves horizontally toward the vertical axis of the nearest robot.
4. If a robot sees robots both below on its vertical axis and on its right, then it computes a certain destination point and performs a diagonal move down and to the right.

It is proved that, assuming the initial visibility graph is connected, the robots rendezvous after a finite number of events. For example, starting as in the figure before, the lower-right robot will not move, and the other three will become collocated with it.

Fig. 1 Synchronized
compasses

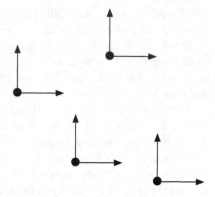

The proof is quite complicated, because, although each robot goes through a sequence
of event cycles Wait-Look-Compute-Move, the robots are entirely unsynchronized,
and so a robot may start to move before another has finished moving.

More General References

More general references for the subject of distributed robotics are the books [6, 28,
33, 35], and the survey [34].

References

1. Abdessameud, A., Tayebi, A.: Global trajectory tracking control of VTOL-UAVs without linear velocity measurements. Automatica **46**(6), 1053–1059 (2010)
2. Ando, H., Oasa, Y., Suzuki, I., Yamashita, M.: Distributed memoryless point convergence algorithm for mobile robots with limited visibility. IEEE Trans. Robot. Autom. **15**, 818–828 (1999)
3. Bespamyatnikh, S., Bhattacharya, B., Kirkpatrick, D., Segal, M.: Mobile facility location. In: Fourth International ACM Workshop on Discrete Algorithms and Methods for Mobile Computing and Communications, pp. 46–53 (2000)
4. Blondel, V.D., Hendrickx, J.M., Olshevsky, A., Tsitsiklis, J.N.: Convergence in multiagent coordination, consensus, and flocking. In: Proceedings of IEEE Conference on Decision and Control and European Control Conference, pp. 2996–3000 (2005)
5. Brockett, R.W.: Asymptotic stability and feedback stabilization. In: Brockett, R., Millman, R., Sussmann, H. (eds.) Differential Geometric Control Theory, pp. 181–191. Birkhauser (1983)
6. Bullo, F., Cortés, J., Martínez, S.: Distributed Control of Robotic Networks. Princeton University Press, Princeton (2009)
7. Castillo, P., Dzul, A., Lozano, R.: Real-time stabilization and tracking of a four-rotor mini rotorcraft. IEEE Trans. Control Syst. Technol. **12**(4), 510–516 (2004)
8. DeGroot, M.H.: Reaching a consensus. J. Am. Stat. Assoc. **69**(342), 118–121 (1974)
9. do Carmo, M.P.: Differential Geometry of Curves and Surfaces. Prentice-hall, Englewood Cliffs (1976)
10. Dörfler, F., Bullo, F.: On the critical coupling for Kuramoto oscillators. SIAM J. Appl. Dyn. Syst. **10**(3), 1070–1099 (2011)
11. Dörfler, F., Bullo, F.: Synchronization and transient stability in power networks and nonuniform Kuramoto oscillators. SIAM J. Control Optim. **50**(3), 1616–1642 (2012)
12. Dörfler, F., Chertkov, M., Bullo, F.: Synchronization in complex oscillator networks and smart grids. Proc. Natl. Acad. Sci. **110**(6), 2005–2010 (2013)
13. Feintuch, A., Francis, B.: Infinite chains of kinematic points. Automatica **48**, 901–908 (2012)
14. Flocchini, P., Prencipe, G., Santoro, N., Widmayer, P.: Gathering of asynchronous robots with limited visibility. Lect. Notes Comput. Sci. **2010**, 247–258 (2001)
15. Hua, M.D., Hamel, T., Morin, P., Samson, C.: Introduction to feedback control of underactuated VTOL vehicles: a review of basic control design ideas and principles. IEEE Control Syst. **33**(1), 61–75 (2013)

© The Author(s) 2016
B.A. Francis and M. Maggiore, *Flocking and Rendezvous in Distributed Robotics*, SpringerBriefs in Control, Automation and Robotics, DOI 10.1007/978-3-319-24729-8

16. Jadbabaie, A., Lin, J., Morse, A.S.: Coordination of groups of mobile autonomous agents using nearest neighbor rules. IEEE Trans. Autom. Control **48**(6), 988–1001 (2003)

17. Jadbabaie, A., Motee, N., Barahona, M.: On the stability of the Kuramoto model of coupled nonlinear oscillators. In: Proceedings of the American Control Conference, pp. 4296–4301 (2004)

18. Justh, E., Krishnaprasad, P.S.: Steering laws and continuum models for planar formations. In: Proceedings of CDC (2003)

19. Khalil, H.K.: Nonlinear Systems, 2nd edn. Prentice-Hall, Upper Saddle River (2002)

20. Krick, L., Broucke, M.E., Francis, B.A.: Stabilization of infinitesimally rigid formations of multirobot networks. Int. J. Control **82**, 423–439 (2009)

21. Kuramoto, Y.: Self-entrainment of a population of coupled nonlinear oscillators. In: Araki, H. (ed.) International Symposium on Mathematical Problems in Theoretical Physics, Lecture Notes in Physics, vol. 39, p. 420. Springer, Berlin (1975)

22. Lee, T., Leok, M., McClamroch, N.H.: Nonlinear robust tracking control of a quadrotor UAV on SE(3). Asian J. Control **15**(2), 391–408 (2013)

23. Lin, Z.: Coupled dynamic systems: From structure towards stability and stabilizability. Ph.D. thesis, University of Toronto (2006)

24. Lin, Z., Broucke, M., Francis, B.: Local control strategies for groups of mobile autonomous agents. IEEE Trans. Autom. Control **49**, 622–629 (2004)

25. Lin, Z., Francis, B.A., Maggiore, M.: Necessary and sufficient graphical conndditions for formation control of unicycles. IEEE Trans. Autom. Control **50**, 121–127 (2005)

26. Lin, Z., Francis, B.A., Maggiore, M.: State agreement for coupled nonlinear systems with time-varying interaction. SIAM J. Control Optim. **46**, 288–307 (2007)

27. Marshall, J.A., Broucke, M.E., Francis, B.A.: Formations of vehicles in cyclic pursuit. IEEE Trans. Autom. Control **49**, 1963–1974 (2004)

28. Mesbahi, M., Egerstedt, M.: Graph Theoretic Methods in Multiagent Networks. Princeton University Press, Princeton (2010)

29. Moreau, L.: Stability of multi-agent systems with time-dependent communication links. IEEE Trans. Autom. Control **50**, 169–182 (2005)

30. Nair, S., Leonard, N.E.: Stable synchronization of mechanical system networks. SIAM J. Control Optim. **47**(2), 661–683 (2008)

31. Ren, W.: On consensus algorithms for double-integrator dynamics. IEEE Trans. Autom. Control **53**(6), 1503–1509 (2008)

32. Ren, W., Atkins, E.: Distributed multi-vehicle coordinated control via local information exchange. Int. J. Robust Nonlinear Control **17**(10–11), 1002–1033 (2007)

33. Ren, W., Beard, R.: Distributed Consensus in Multi-vehicle Cooperative Control. Communication and Control Engineering Series. Springer, London (2008)

34. Ren, W., Beard, R.W.: A survey of consensus problems in mulit-agent coordination. In: Proceedings American Control Conference, pp. 1859–1864 (2005)

35. Ren, W., Cao, Y.: Distributed Coordination of Multi-agent Networks. Communication and Control Engineering Series. Springer, London (2011)

36. Reynolds, C.W.: Flocks, herds and schools: a distributed behavioral model. In: Proceedings of the 14th Annual Conference on Computer Graphics and Interactive Techniques (SIGGRAPH '87), vol. 21, pp. 25–34 (1987)

37. Roza, A., Maggiore, M., Scardovi, L.: A class of rendezvous controllers for underactuated thrust-propelled rigid bodies. In: Proceedings of the IEEE Conference on Decision and Control (CDC), pp. 1649–1654 (2014)

38. Sarlette, A., Sepulchre, R., Leonard, N.E.: Autonomous rigid body attitude synchronization. Automatica **45**(2), 572–577 (2009)

39. Scardovi, L., Sarlette, A., Sepulchre, R.: Synchronization and balancing on the N-torus. Syst. Control Lett. **56**, 335–341 (2007)

40. Sepulchre, R., Paley, D.A., Leonard, N.E.: Stabilization of planar collective motion: all-to-all communication. IEEE Trans. Autom. Control **52**(5), 811–824 (2007)

41. Strogatz, S.H.: From Kuramoto to Crawford: Exploring the onset of synchronization in populations of coupled oscillators. Physica D **143**, 1–20 (2000)
42. Suzuki, Yamashita: Distributed anonymous mobile robots:formation of geometric patterns. SIAM J. Comput. **28**(4), 1347–1363 (1999)
43. Swaroop, D., Hedrick, J.K.: String stability of interconnected systems. IEEE Trans. Autom. Control **41**(3), 349–357 (1996)
44. Tanner, H.G., Jadbabaie, A., Pappas, G.J.: Flocking in fixed and switching networks. IEEE Trans. Autom. Control **52**(5), 863–868 (2007)
45. Tsitsiklis, J.N., Bertsekas, D.P., Athans, M.: Distributed asynchronous deterministic and stochastic gradient optimization algorithms. IEEE Trans. Autom. Control **31**(9), 803–812 (1986)
46. Vicsek, T., Czirók, A., Ben-Jacob, E., Cohen, I., Shochet, O.: Novel type of phase transitions in a system of self-driven particles. Phys. Rev. Lett. **75**, 1226–1229 (1995)

Printed in the United States
By Bookmasters